Starting Work

Electrical Installation Series – Foundation Course

Ted Stocks
Thomas Awcock
Terry Brown

Edited by Chris Cox

Starting Work

Copyright © CT Projects 1998

The Thomson logo is a registered trademark used herein under licence.

For more information, contact Thomson Learning, High Holborn House; 50-51 Bedford Row, London WC1R 4LR or visit us on the World Wide Web at:
http://www.thomsonlearning.co.uk

British Library Cataloguing-in-Publication Data
A catalogue record for this book is available from the British Library

---

ISBN 1-86152-758-6

First published 1998 by Macmillan Press Ltd
Reprinted 2002, 2003 and 2004 by Thomson Learning

Printed in Croatia by Zrinski d.d.

About this book

"Starting Work" is one of a series of books published by Thomson Learning related to Electrical Installation Work. The series may be used to form part of a recognised course, for example City and Guilds Course 2360, or individual books can be used to update knowledge within particular subject areas. A complete list of titles in the series is given below.

Foundation Course books give the student the underpinning knowledge criteria required for City and Guilds Course 2360 Part I Theory. The supplementary book, Practical Requirements and Exercises, covers the additional underpinning knowledge required for the Part I Practice.

Level 2 NVQ

Candidates who successfully complete assignments towards the City and Guilds 2360 Theory and/or Practice Part I can apply this success towards Level 2 NVQ through a process of Accreditation of Prior Learning.

Electrical Installation Series

Foundation Course
Starting Work
Procedures
Basic Science and Electronics

Supplementary title:
Practical Requirements and Exercises

Intermediate Course
The Importance of Quality
Stage 1 Design
Intermediate Science and Theory

Supplementary title:
Practical Tasks

Advanced Course
Advanced Science
Stage 2 Design
Electrical Machines
Lighting Systems
Supplying Installations

Acknowledgements

The authors and publishers gratefully acknowledge the following illustration sources:

Abru Ltd for Figure 9.1; AVO International for Figure 10.2; Balcan Engineering Ltd for Figure 5.18; Clipsal UK Limited for Figure 3.1; F8 Imaging for Figures 2.2, 2.3, 5.1, 5.2; Maplin Electronics for Figures 1.1, 1.2, 1..3, 2.5; RS Components for Figures 3.2, 3.24, 3.25, 3.31, 4.1, 4.2, 4.3, 6.1, 6.2, 7.1, 7.2, 8.1, 8.2, 8.3, 8.4, 9.21, 10.1; Stocksigns Ltd for Figures 4.11–4.26; The National Grid Company plc for Figures 1.16, 1.18, 1.19; Sandra Truscott for Figure 2.1.

Every effort has been made to trace all copyright holders but if any have been inadvertently overlooked, the publishers will be pleased to make the necessary arrangements at the first opportunity.

Study guide

This studybook has been written to enable you to study either in a classroom or in an open or distance learning situation. To ensure that you gain the maximum benefit from the material you will find prompts all the way through that are designed to keep you involved with the subject. The book has been divided into 24 parts each of which may be suitable as one lesson in the classroom situation. However if you are studying by yourself the following points may help you.

☞ Work out when, and for how long, you can study each week. Complete the table below and from this produce a programme so that you will know approximately when you should complete each chapter and take the progress and end tests. Your tutor may be able to help you with this. It may be necessary to reassess this timetable from time to time according to your situation.

☞ Try not to take on too much studying at a time. Limit yourself to between 1 hour and 2 hours and finish with a task or the self assessment questions (SAQ). When you resume your study go over this same piece of work before you start a new topic.

☞ You will find the answers to the questions at the back of this book but before you look at the answers check that you have read and understood the question and written the answer you intended.

☞ A "progress check" at the end of Chapter 5 and an "end test" covering all the material in this book are included so that you can assess your progress.

☞ Tasks are included where you are given the opportunity to ask colleagues at work or your tutor at college questions about practical aspects of the subject. There are also tasks where you may be required to use manufacturers' catalogues to look answers up. These are all important and will aid your understanding of the subject.

☞ You will need to have available for reference a current copy of IEE Guidance Note 3 in order to complete some of the exercises in Chapter 10.

☞ Your safety is of paramount importance. You are expected to adhere at all times to current regulations, recommendations and guidelines for health and safety.

Study times					
	a.m. from	to	p.m. from	to	Total
Monday					
Tuesday					
Wednesday					
Thursday					
Friday					
Saturday					
Sunday					

Programme	Date to be achieved by
Chapter 1	
Chapter 2	
Chapter 3	
Chapter 4	
Chapter 5	
Progress check	
Chapter 6	
Chapter 7	
Chapter 8	
Chapter 9	
Chapter 10	
End test	

Contents

1

Electricity in the Environment

You will need to have available for reference a current copy of IEE Guidance Note 3 in order to complete some of the exercises in Chapter 10. You will be prompted at the beginning of that chapter so that you can obtain the relevant material before you start.

At the beginning of all the other chapters in this book you will be asked to complete a revision exercise based on the previous chapter – Sid with a clipboard will remind you of this. For the first exercise add to the list below any **labour-saving** electrical equipment that you can find in and around the home and at your place of work.

washing machine
electric drill
electric carving knife

Figure 1.1   Audio and video equipment.

Figure 1.2   Cable and accessories.

On completion of this chapter you should be able to:

◆ state the benefits gained from the utilisation of electricity and electricity-operated devices at home, at work and in leisure activities
◆ state the different types of power station in use
◆ draw a graph of the consumption of electrical energy against time
◆ list the requirements for the siting of a coal-fired power station
◆ identify the relationships of generated output of each of the types of power stations stated
◆ state the standard generation and transmission voltages in use in the UK
◆ explain why it is important to have a network grid system
◆ complete the revision exercise at the beginning of the next chapter

Figure 1.3   Mains electric and lighting equipment.

Part 1

The use of electricity

Although infrequent, it is only when a power cut occurs that we realise just how much we rely on our electricity supply. Just think about some of the electrical equipment we take for granted, and the effects that these have on our life.

At home electricity is used to power lights, cookers, heaters, televisions, radios, CD players, kettles, toasters, washing machines... the list seems endless. Don't forget that battery-operated items also rely on electricity. All these items contribute not only to our comfort but to our safety and security as well. These items include smoke detectors, security lighting, fire alarms and time clocks for central heating.

You may also find many electrical items at your place of work. Many electrical contractors will use a computer to keep customers' records and to help with design and larger firms may have several computers all linked into a network. Many of the tools used at work will be electrically operated, for example angle grinders, drills and jigsaws, and machinery is often driven by motors. At home and at work many of the electrical, and electrically operated devices, save us work – they are "labour-saving".

We use electricity for communications and storing data; telephones, answering machines and faxes all require electricity. At work we use computers to store data such as stock checks, contacts, business letters and orders.

When you finish work you may enjoy going dancing or down to the local sports centre, but if there was a prolonged power cut neither would be able to function!

In locations such as hospital operating theatres, where it is imperative that the supply continues to function even if the power is cut, standby generators are installed.

The constant supply

Whenever we switch on electrical equipment we expect the supply to be there. Mains electricity is not stored, it is generated all the time. This means that electrical generators must be working 24 hours a day, every day of the year. To make sure there is enough power available for any anticipated load, generators over the whole country have to be running regardless of whether the supply is used.

A typical demand for electrical power over a 24 hour period is shown in Figure 1.4. You can see from the graph that there are times of the day when demand hits very high peaks and times at night where demand is low. Generating companies have computerised systems so that the amount of power being generated will meet the demands for each time of day.

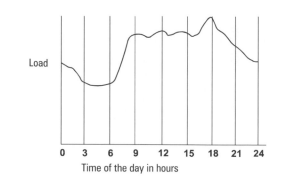

Figure 1.4   The load demand over 24 hours.

If a line is drawn through the curve as shown in Figure 1.5 and this is taken as being the average output from the country's generators, then there would be times of above-average demand and times below. During the periods when the demand is below the average supply, generators would be running but not all of their output would be used. This means that during these times the generating companies could sell the electricity more cheaply in order to recoup some of their generation costs.

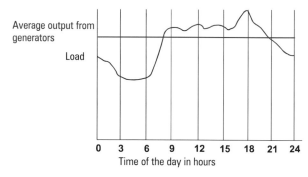

Figure 1.5

When the demand is above the average the supply companies would have to run up more generators to cope. If we assume that the most economical power stations are in use to meet the average demand, then the fast run-up, less economical generators would be used to supply the high peak loads. This would mean that the above average supply would be more expensive to produce.

Although the methods used by the generating companies are much more sophisticated than this, the theory is similar.

Remember
Mains electricity is not stored, it is generated all the time.

One day when you are at home plot a curve for the load taken over a 12 hour period. This can be done at the metering point in your home by recording the readings in each hour throughout a single day and writing them on the table.

You should only need to use the units and 1/10th units readings, but you will need to determine your own scale depending on the amount of electricity you use.

Results Table

Time	Reading	Amount of electricity used.
08.00		
09.00		
10.00		
11.00		
12.00		
13.00		
14.00		
15.00		
16.00		
17.00		
18.00		
19.00		
20.00		

Amount of electricity used each hour

8 to 9	9 to 10	10 to 11	11 to 12	12 to 13	13 to 14	14 to 15	15 to 16	16 to 17	17 to 18	18 to 19	19 to 20

Time of day

Power stations

Electricity is generated by rotating the shaft of an alternator. The energy required to drive an alternator can come from several sources, which can be divided into three categories:

- steam plant
- direct drive systems
- natural resources

Steam plant

Most electricity is generated by producing heat through an energy conversion process. The heat is then used to change water into steam and the steam drives turbines which in turn rotate the alternator. Figure 1.6.

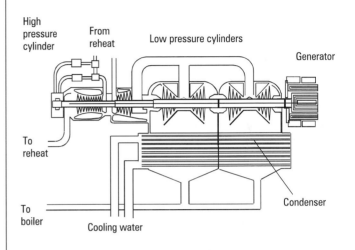

Figure 1.6   Steam plant.

Coal has been used more than any other fuel to produce the heat in power stations. A station with an output of 2000 MW (2 thousand million watts) consumes about 5 million tonnes of coal a year. For this reason coal-fired power stations (Figure 1.7) are often sited near to coal mines or on easy transport routes, such as road, rail or sea.

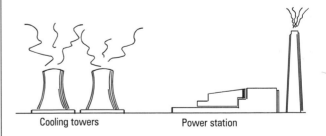

Figure 1.7   Coal-fired power station.

It is necessary to have high temperatures to create the steam, but it is also necessary to cool parts of the system down. Where possible power stations are situated near to a large source of water, such as the sea or a large river. A 2000 MW station requires about 60m tonnes of cooling water each second.

Where the large quantities of water required are not available the cooling process has to be helped by cooling towers. Within a cooling tower about one-third of the heat contained in the cooling water is transferred to air passing up the tower; the remaining two-thirds is absorbed as latent heat, evaporating about one per cent of the water flow. This water vapour also passes up the tower and is discharged into the atmosphere. These towers can often be seen producing large amounts of water vapour and are sometimes confused with chimneys (Figure 1.7).

Nuclear power stations (Figure 1.8) also use the steam process. The siting of these stations relates more to possible hazards from the nuclear fuel than from its transportation. In comparison with coal the amount of fuel used is very small – about 4.5 tonnes of uranium each week for a 1000 MW station.

The cooling, however, still has the same requirements as coal stations, and the disposal of the waste radioactive materials is an additional problem.

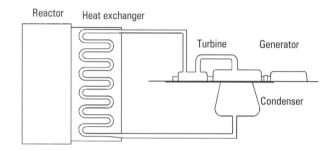

Figure 1.8   Nuclear power station.

Oil can also be used as the heating fuel in a steam plant. This can be either as a refined fuel or oil in its crude state. Calling it "refined" is not strictly true, for it is the residue, after the more valuable parts have been taken away, that is used in power stations. This type of station would normally be sited close to an oil refinery so that transportation is not a problem.

Oil can be burnt in its crude state; however, pumping it through pipes over long distances can create problems, as the pipes would need to be heated to keep the oil flowing. For this reason power stations using crude oil are usually sited close to deep water berths (Figure 1.9) so that tankers can deliver the oil by sea. Some coal-fired power stations are designed to use oil in emergencies.

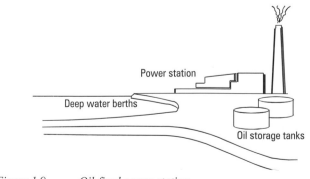

Figure 1.9   Oil-fired power station.

Direct drive

These have been identified separately as the alternators are driven directly from a machine and not through a separate process like steam.

Gas-turbine generators (Figure 1.10) are used so as to be run up when there is extra demand for a short period of time. They are basically an aircraft "jet" engine coupled to an alternator. They can be run up to speed very fast when required. However, their efficiency is very low so they are not used more than they need to be.

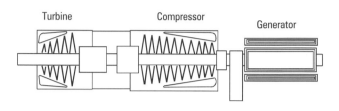

Figure 1.10   Gas-turbine generator.

Diesel engines (Figure 1.11) are also used to drive alternators. These are used extensively for standby supplies in factories and shops, and are designed to cut in automatically in the event of a main supply failure. They are also used for supplying remote areas where it is uneconomical, or very difficult, to install the mains grid systems. Islands remote from the mainland supplies often have their own diesel generator stations.

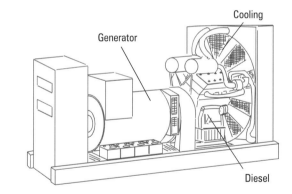

Figure 1.11   Diesel generator.

Natural resources

The most widely used natural resource for generating electricity is **hydro-electric** power. In some countries where lakes and reservoirs are plentiful the electricity supplies are almost entirely generated by hydro-electric plants. Although electricity is cheap to produce in this way it often has to be transmitted over long distances to the nearest populated and industrial areas. To produce electricity using hydro-electric power a head of water has to be available. Figure 1.12 shows how reservoirs high in the mountains are used to create pressure through an artificial tunnel. By controlling the flow in the tunnel the output can be varied throughout the day. In areas where the rainfall is not consistent, "cheap" electricity, produced by steam and nuclear generators, can be used to pump the water back into the top reservoir during the night. Then during the day the hydro-generator can be used to produce electricity at peak times.

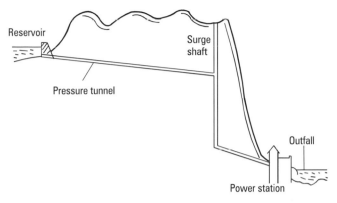

Figure 1.12   Hydro-electric station.

Tidal flow is used in some countries as a means of generating electricity. Where there are large tidal changes, a dam is built and a hydro-electric system installed. The comparatively low head of water in this system means that it is not very efficient. However, once it is built there are no fuel costs, so the electricity generated is very cheap. The main disadvantages are that the tides are different every day and therefore the electricity is often not being generated at times of peak demand.

Wind generators (Figure 1.13) are installed in many countries throughout the world. The output of each generator is very small compared with a power station and is completely dependent on the wind blowing at a consistant speed.

Figure 1.13   Wind generator.

Geothermal stations are now being developed which use the very high temperatures from the earth's core to produce steam and drive generators.

The electricity supply in the United Kingdom comes from a variety of sources. Table 1.1 illustrates the approximate provision as a percentage of total energy supplied.

Table 1.1

The generation of electricity in the UK	
Coal	42%
Fuel oil	4.5%
Nuclear	28%
Natural gas	21%
Other	1.5%
Imports	2.0%

Points to remember ◄ – – – – – – – – – – – – – – –

Power stations using crude oil would be situated near to a deep water dock. Why?

What is discharged from the top of a cooling tower?

Where would you find a hydro-electric plant situated?

What consideration is given to the siting of nuclear power stations?

Give an example of a use of electricity in the following situations?
data storing	*computer*
communication	. . . .
comfort	. . . .
safety	. . . .
security	. . . .
entertainment	. . . .
labour-saving (work)	. . . .
labour-saving (home)	. . . .

Part 2

Electricity generation and the grid system

Generation to transmission

Having seen how electricity is generated we can now look at how it is distributed to where it is required.

The output voltage of a generator set in a power station will not be greater than 25 000 volts a.c., and on older generators it may be considerably less.

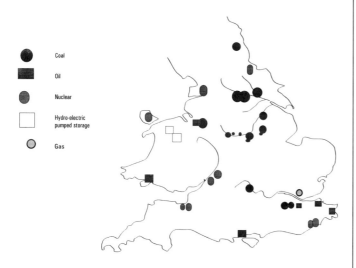

Figure 1.14   *The siting of major power stations in England and Wales.*
Reproduced with kind permission of NATIONAL GRID.

For transmission purposes the generated output must be increased to a maximum of 400 000 volts (400 kV) or 275 000 volts (275 kV) in some situations.

The companies responsible for the generation of electricity include National Power and PowerGen. The transmission company is the National Grid Company (NGC) and it is jointly owned, through a holding company, by the distribution companies.

Transmission switching stations

When the electricity comes from the generator it passes through a switching station which transforms it up to the transmission voltages.

The switching station, as the name implies, is more than just a transformer. As with all circuits there must be means of overcurrent protection and means of isolation. A circuit breaker provides the means of overcurrent protection, and is designed to operate in a fraction of a second under overcurrent conditions. The isolators are not designed for switching when the load is connected, and can only be used when the circuit breaker has already disconnected the supply. As can be seen from Figure 1.15, there are two isolators.

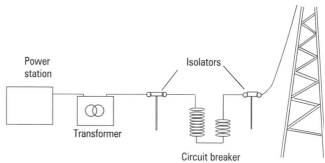

Figure 1.15

These can be used so that the circuit breaker can be completely isolated from the grid or the transformer, both of which may be "alive". This enables the circuit breaker to be maintained without any danger from the supply.

Transmission

The majority of electrical power is transmitted on overhead cables suspended from pylons (Figure 1.16). Although the pylons can look unsightly it is by far the most cost-effective way of transmitting power.

High voltages are used for transmission and these have to be well insulated from one another and from earth.

When bare conductors are placed high in the air (Figure 1.17), and spaced apart, then the space between the conductors and the conductors and earth becomes the insulation and no further covering is required. The air around the conductors also helps to keep them cool when they are carrying heavy loads.

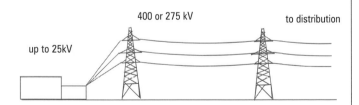

Cables are bunched in twos
or fours as shown

Figure 1.16   Reproduced with kind permission of
NATIONAL GRID.

Figure 1.18   Reproduced with kind permission of
NATIONAL GRID.

Figure 1.18 represents a 400 kV oil-filled cable of 1936 square millimetres copper cross-section showing the lapped construction. This cable could be used for underground transmission and its overall diameter would be 126 millimetres (about $\frac{2}{3}$rds of the width of this page – check with your ruler).

The transmission system

When the 132 kV transmission network was first established it was known as the Grid System (Figure 1.19). As demand grew the 275 kV Supergrid System was constructed and we now have an integrated 400/275 kV main transmission system. This consists of a latticework of cables which connect power stations and large load areas together.

400 or 275 kV

up to 25kV

to distribution

Figure 1.17   Transmission voltages.

To install high-voltage transmission cables underground is very costly. The cables need special cooling methods, such as oil being pumped through them, as the natural heat dissipation into the surrounding soil is not sufficient. They also require very thick insulation to protect the conductors from each other and the earth around them. If a fault develops on the cables when they are exposed in the air, this can be seen and often it is a comparatively straightforward job to put it right. However, if a fault develops on an underground cable this is not readily accessible and can take time and cost far more to repair. Having said this, 400 kV cables are placed underground when there are no alternatives or it is important for environmental reasons, and in the UK there are more of these cables underground than anywhere else in the world.

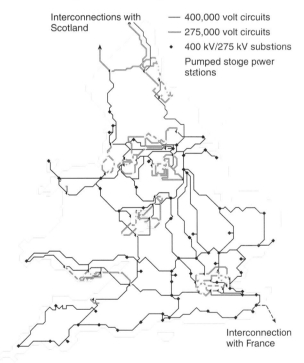

Interconnections with Scotland

— 400,000 volt circuits
— 275,000 volt circuits
♦ 400 kV/275 kV substions
Pumped stoage power stations

Interconnection with France

Figure 1.19   The grid system in England and Wales.

For clarity only the main lines have been shown.
Reproduced with kind permission of
NATIONAL GRID.

Try this

Complete the following:

The output voltage of a generator set in a power station will not be more than _____ and will be stepped up to a maximum of _____ for transmission.

Earlier in this chapter we established that power stations are not always situated where the highest loads are. This means that there must be cables from the power stations to all of the load points. There must also be connections between power stations so that they can cover for each other.

Nationally all peak loads do not occur at the same time and a power station in one area may be called on to supply power to another. From time to time power stations have to be closed down for maintenance and repair. At these times other power stations must meet the total demand.

Over the last 30 years the trend has been to build bigger and more efficient power stations and close down the smaller uneconomical ones. This has meant a reduction from 300 to less than half that number, and as a consequence the links between the major new stations have become more important.

In addition to the transmission system in the UK there are cross-channel links with the French supply system. These are underwater cables going from the Kent coast to North East France. The time and lifestyle differences in the two countries are factors which make these links practicable. The peak demand times vary, and this allows the UK to import power from France when demand requires. It also allows the export of energy to take place when the French demand is high. For practical reasons these cross channel cables are supplied with d.c. and this is converted to a.c. at each end.

Remember

Electricity is generated as alternating current.
This means that it can be transformed up or down for transmission and distributed to where it is required.

Points to remember ◀ – – – – – – – – – – – – –

What is the maximum output voltage of a generator set in a power station?

What provides the means of overcurrent protection at a switching station?

How many are there and why?

Which is the most cost-effective way of transmitting electrical power?

In environmentally sensitive regions the above method may not be suitable. Describe the method for transmitting electrical power under these circumstances.

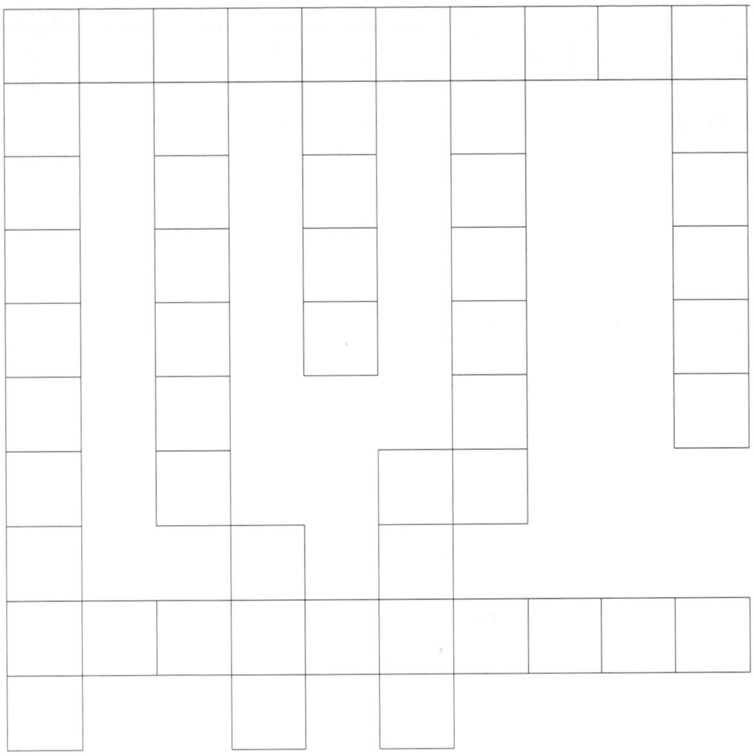

The following words will all fit together in the grid above. See if you can find how they fit.

SUPPLY

ALTERNATOR

GENERATORS

SEA

TURBINE

GEOTHERMAL

WIND

ROTOR

NUCLEAR

Part 3

Distribution to the consumer

Distribution

Although systems of distribution can sometimes appear to be very complex, they are far better than was available in the past. For example, in London in 1919 there were eighty separate supply undertakings with seventy different generating stations, fifty different supply systems operating at twenty four different voltages and ten different frequencies. Since those times voltages and frequencies have been standardised (Figure 1.20).

There are now 12 regional boards in England and Wales who are distribution companies responsible for the supply of electricity to the consumers in their area. Each of these companies purchases its electricity from a generating company, National Power or PowerGen. The power is distributed by the National Grid Company which is jointly owned by the 12 Regional Electricity Companies. Scotland and Northern Ireland have their own electricity supply systems.

The Regional Electricity Companies have a legal responsibility to keep the supply within certain limits. Following voltage standardisation in Europe these are, for voltage, a nominal supply of 400/230 V, +10%, –6%; and for frequency it must not be more or less than 1% of 50 Hz during a 24 hour period.

Try this

1. If the Electricity Company's stated voltage is 230 V, what are the limits they must keep within?
   (a) the supply's highest voltage

   (b) the supply's lowest voltage

2. If the frequency is stated to be 50 Hz what are the limits the Electricity Company must keep within?
   (a) highest frequency

   (b) lowest frequency

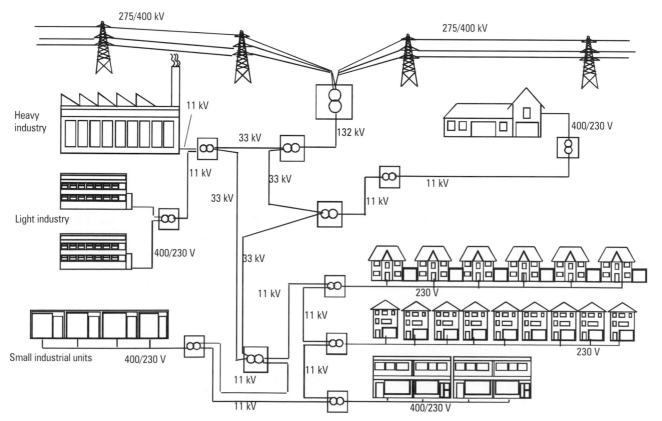

Figure 1.20   Distribution system showing the voltages used.

The distribution system can be split into three main sections
- industrial
- commercial and domestic
- rural

The reason for splitting these is the different voltages they require and the remoteness of rural supplies. Wherever possible, distribution cables are laid underground. Apart from rural areas, almost all 11 kV and 400/230 V cables are buried underground, and a significant number of the higher voltage cables are now also buried.

The main intake

After the supply has been generated and distributed it will eventually end up at the consumer's main supply intake.

Regardless of the size of the installation, there are several items of equipment that must be included at every main intake position. The first of these is an overcurrent device protecting the supply company's cables. This is the property of the company and is sealed so that it cannot be tampered with. On some very large installations this protection may be incorporated in the main control panel, but it would still be sealed by the supply company. All installations are metered so that payment for energy used can be calculated. The energy or kilowatt hour (kWh) meter, as it is often referred to, must be connected between the supply and any of the consumer's equipment. This ensures that the supply company charges for *all* the electricity used.

It is after the energy meter that the consumer's installation begins. The first item of equipment that the consumer must have is a method of switching the installation off. Next there must be protection devices for disconnecting each circuit in the event of a fault developing. These may often be included in one piece of equipment called a consumer unit (Figure 1.21).

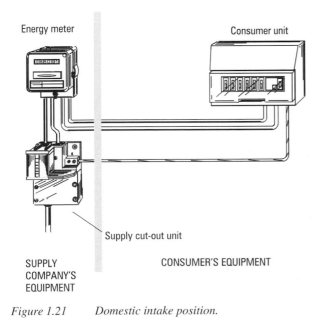

Figure 1.21   Domestic intake position.

The connection to earth

In the event of a fault occuring between a live conductor and exposed metalwork, enough current has to flow through the conductor to make the protection device operate, Figure 1.22 shows the path that a fault would take. The effect is to short the transformer out so that high currents are drawn from the system. If the protection device does not operate very fast the energy used may cause the cables to overheat and melt, and a fire to start.

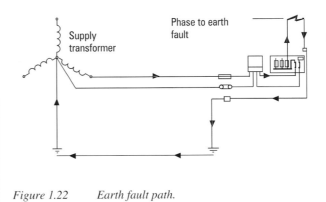

Figure 1.22   Earth fault path.

Points to remember ◄ – – – – – – – – – – – – – – –

It is only when we have power cuts that we really appreciate the benefits that we gain from using electricity. At home, at work or in our leisure time electricity provided to us gives us greater comforts and contributes to our health and security and enables us to use labour-saving equipment.

Mains electricity cannot be stored, but must be generated continually. The fact that we do not use the same quantity of electrical power all the time means that there are times of peak and low demand.

Most electricity in the UK is generated by burning _____ to produce heat, which produces _____ and the _____ drives _____ connected to the generators. Nuclear and oil-type power stations also use this steam process. An alternative method is where generators are driven directly from motors, such as diesel or jet engines.

By far the cheapest method of producing electricity is by _____ power, but this is restricted to mountainous areas with high rainfall. Other systems using natural resources are being developed, but as yet have very small outputs. Examples are _____ _____.

After the electricity has been generated it has to be transmitted around the country. It then has to be distributed to the consumers for their use. A grid system of transmission is used to give an economic, efficient supply to each area.

Which organisation is responsible for electricity generation?
_____

Which organisation is responsible for national electricity transmission? _____

Which organisation is responsible for local electricity distribution? _____

The output voltage of a generator set in a power station will not be greater than _____.

The system of high-voltage transmission is known as
_____.

What is the maximum voltage for transmission?
_____

The supply to heavy industry plants is likely to be at
_____.

The supply to small industrial units is likely to be at
_____

The nominal supply voltage to a residential house is likely to be at _____

Name two pieces of supply company's equipment that must be included at every intake position.
_____ and _____

Where does the consumer's installation begin?
_____

What is another name for a kilowatt hour meter?
_____

What is the voltage supply if it is 5% greater than the nominal supply voltage of 230 V? _____

Fill in the missing voltages after the transmission voltage has been reduced by the step-down transformers.

    275/400 kV

    _____

    _____

    _____

    400/230 V

2
Working Relationships

Before we continue with this chapter see if you can remember some important points from Chapter 1.

Which is the most widely used natural resource for generating electricity?

Name three types of steam plant used to generate electricity.

What voltages are likely to be supplied to
heavy industry _____?
light industry _____?
domestic property _____?

Figure 2.1   *Answering the telephone.*

Figure 2.2   *Working on site.*

On completion of this chapter you should be able to:

◆ describe the structure of a typical electrical contracting company
◆ state the role of the contracting team including the client
◆ state the need to represent your organisation favourably and how to achieve this
◆ identify the main requirements for achieving and maintaining good working relationships with customers and co-contractors
◆ be able to identify customers' requirements both in person and by telephone
◆ state the necessity of formal contract procedures
◆ be able to establish and maintain "on site" relationships with visitors
◆ complete the revision exercise at the beginning of the following chapter

Figure 2.3   *Working in the workshop.*

Part 1

The team

A business cannot exist without its customers! So the success or failure of an organisation can be determined by the way it is represented by its staff (Figure 2.4). Our aim should be to provide a caring, polite and professional image and deliver a first-class level of service.

Figure 2.4   Oops!

Part of a team

When you work for an electrical contracting company you are part of a team. If it is a large company there may well be staff in all these areas:

- general management
- works management
- drawing office
- estimating
- personnel
- safety
- clerical
- stores
- transport
- approved electricians
- trainees

If the company is smaller most of these areas will be covered by fewer people.

Remember
Ultimately the customer pays the wages. Tidiness and a polite manner costs nothing.

Do you know what happens in your company before you are sent to a job?

Let us consider a contract for a project such as a new factory installation. The same principles apply for most projects, even the simple additional socket outlet in a domestic installation. In the case of smaller contractors all the functions may be carried out by one person.

Generally, clients will ask for a price for the work they have in mind. The client will give the company all the necessary details regarding the job. The design staff will then draw up a specification and it will be sent to the estimating staff for pricing. The clerical staff will type the quote to be sent to the client. If the quote is accepted then the stores' staff will arrange to have all the necessary goods and equipment in stock at the required time for the electricians to do the job. On some larger jobs transport will have to be made available.

Whatever size your company is, success is dependent upon each of the activities and all members of the team.

First impressions

First impressions are most often the ones by which a company or its individual staff are judged. Of course this judgement will depend on circumstances such as the location and conditions of work but generally the same rules apply.

Are the people (members of staff):
- as clean and tidy as possible?
- polite and helpful?
- knowledgeable?
- on time for appointments?

Do they:
- treat the customer's property with care?
- take the right equipment to do the job and have the correct tools?

Is the:
- company vehicle clean and well kept?
- office and workshop tidy and clean?

Are:
- accurate records kept and are they readily available?
- customers made welcome and not kept waiting (when they visit the company's premises)?

A positive response in all of these areas means that a customer may be encouraged to do business with the company, recommend the company to others and return for repeat business.

The telephone

The telephone is frequently the first point of customer contact. A customer will judge the company on the way that the call is handled.

So it is important that calls are answered promptly.

If you answer the company telephone it is important that:

- callers are answered politely, they are correctly identified and their requirements accurately established
- the customer is informed of any problems and what action is being taken
- callers are only given information which can be disclosed
- incoming and outgoing calls are dealt with politely *throughout* the call
- the purpose of an outgoing call is clearly conveyed
- identification (of both company and contact person) is given when making a call
- you announce the name of your company and add your own name
- you do not shout but speak clearly and distinctly

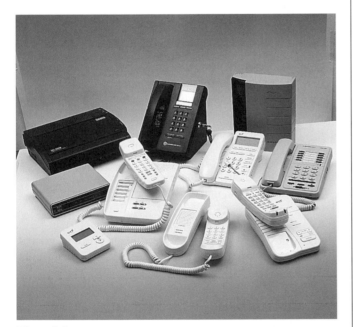

Figure 2.5

If you are required to make a telephone call to a customer or another company it can be helpful if you ensure that you have all the necessary information to hand before you dial the number. You may find that you are connected to an answering machine. If so, wait for the tone then leave your message clearly, identifying both yourself and your company and, if it is relevant, the time and date when the call was made.

Telephone threats

In the event of a telephone threat to the safety of the workplace, such as a bomb alert, terrorist presence or arson attack, the police should be contacted immediately and your manager alerted. If possible you should write down all you can remember.

Try and find out from the caller, and make a note of,

- why there is a problem
- where it is located
- what the problem is

and also make a note of

- any background noise that you can hear,
- whether the caller is male or female,
- if you can detect an accent etc.

All these details may help the police with any investigation.

Try this

Write down as many reasons as you can think of why people might telephone your company.

Put yourself in the customer's shoes and write down your impressions of the last time you telephoned another company.

Messages

Good working relationships are often the result of achieving good communications, whether in person, by telephone or through written messages.

When taking a message make sure that:

- all relevant information is courteously obtained and verified
- messages are always passed on to the correct person/ location
- the information is passed on accurately and clearly
- requests for information are dealt with promptly
- the person is told of any action that may be taken

Points to remember ◀ – – – – – – – – – – – – – –

Name as many job titles of members of staff of an electrical contracting team as you can think of:

A client asks your company for a price for some work. What happens next?

What are the important points regarding answering the telephone for your company?

A new customer telephones your company and wishes to speak to someone with particular specialist knowledge. As this person is not available at present and will have to telephone the customer back, what information should you take from the customer?

Try this

In the space opposite write a letter to your employer reporting the breakage of a fluorescent luminaire on the site where you are working.

Give reasons as to why it happened and request a necessary replacement. State whether the situation requires urgent attention.

Hints:

Keep it brief but to the point.

Where you have a good relationship with your employer, write in an informal and cheerful style. Use a manufacturer's catalogue to quote the precise replacement details: description, make, catalogue no. etc.

As a general rule a letter should start on the top right-hand side with your own address, unless you are using a printed letter heading on behalf of your company. Then you would put the date on the left-hand side followed by the addressee's name and address.

After an introduction, "Dear Sir or Madam or Mr Smith or Miss Taylor" etc., the letter should have a sentence of introduction to the subject matter.

Then start a new paragraph and develop the theme.

Conclude with a short sentence and close with Yours faithfully (if you started Dear Sir or Madam) or Yours sincerely (if you started with Dear Mr Smith or Miss Taylor).

Leave a space for your signature then write your name clearly underneath.

Part 2

Records and reports

Customer records
Where more than one person is likely to deal with a customer a "customer record" must be kept. One should also be kept in a one-person business so that reference to customer details can be made to it when necessary.

The record can be in the form of computer data or a book which is kept in a position available only to staff. Every time contact is made, the details are recorded, including the date, time, action taken and by whom.

For telephone messages companies will often use a special pad (Figure 2.6) located near the telephone which enables you to take down all the relevant details, pass on or file the top copy with the customer's records and leave a duplicate copy in the message book.

TO:	**FROM:**
OF:	
TEL:	**FAX**

Please ring back ◯   **Would like to see you** ◯   **Urgent** ◯

TAKEN BY: **TIME:**   **DATE:**

MESSAGE:

Figure 2.6   *Telephone message sheet.*

Relationships with colleagues
Although it would be desirable, people working together do not always have common aims and attitudes. Colleagues can become jealous over promotion or better pay and conditions. They can also be irritated when they get held up or are unable to work properly through no fault of their own. Therefore it is important that you should ensure that you are on time for work and that you "pull your weight".

You should cooperate with your employer by being punctual and conscientious, both at work and when you are studying, and by following the company rules that apply to you. If you are sick or unable to work through injury you must let your employer know as soon as possible so that different working arrangements can be made.

You should also complete and hand in any paperwork necessary to enable your company to run its business. All records that you complete should be accurate and clearly written so that they can be easily understood by the office staff.

Try this
A colleague has turned up on site and it is obvious from their behaviour that something is wrong. It is important that you work well together. From your own experiences make a note of what the reason/s could be and in what way the situation could be improved.

Records and reports

Most of the information your company requires from you will be on a "**Time Sheet**" (Figure 2.7), but you may also be required to complete **Day Work Sheets** and **Job Sheets** (Figure 2.8), and in addition you may need to make **Reports** appropriate to your work. You should complete these reports in the manner prescribed by your company and at the time specified.

A **Day Work Sheet** is similar to a time sheet but it is filled in on a daily basis and it is generally used to cover extra work that has to be charged for. In addition to the information required on the time sheet it is usual to include the amount of materials that have been used. This is so that the company can charge the customer for both the hours worked and the materials used.

The site foreman or engineer may be required to make frequent **Reports** on the progress of the work on site. This enables any problems encountered to be detected as soon as possible and action to be taken to rectify the situation. Delays can prove to be quite costly.

Variation orders

Variation orders (Figure 2.9), or architect's instructions as they are often called, are issued when any changes in the original order are required. A client who wishes to make an alteration will discuss the change with the architect and then the architect will issue a variation order to the main contractor who will forward it to the subcontractor to whom it applies.

SMITH & SONS   Electrical Contractors

TIME SHEET

Name
Week Ending

	Job No.	Time Started	Time Finished	Total for day	Travelling Time	Mileage and fares
Sun						
Mon						
Tues						
Wed						
Thurs						
Fri						
Sat						
TOTALS						

Foreman's Signature   Date

Figure 2.7   A Time Sheet is completed weekly and includes the hours worked and travelling expenses.

SMITH & SONS   Electrical Contractors

JOB SHEET

Customer

Address

Work to be carried out

Date to carry out work

Special Conditions

Special Instructions

Figure 2.8   A Job Sheet contains the information required to enable the electrician to carry out the work.

Variation Order

Issued by:	George Edward Associates
Employer:	Rachel Louise Hairdressing
Contractor:	Arual Developments Ltd.
Works situated at:	Rachel Louise Hairdressing, Linking Road
Job Ref:	C535
Issue Date:	01-07-9-

Under the terms of the above contract, I/We issue the following instructions:

Instruction:	Office use: £ omit	Office use: £ add
1.1 Add to Contract No. C535, 2 further 13A sockets in the reception area. Quotation from Smith & Sons (Electrical Contractors) attached.		
Signed:		

Amount of Contract Sum	£	
Approx.Value of previous instructions	£	
Approx. Value of this instruction	£	
Approx.adjusted total	£	

Figure 2.9   Variation Order.

Try this

Your company is required to alter an installation. This means that a variation order has been issued. You will need to complete a day work sheet to cover the extra work and materials involved.

Complete the form below (Figure 2.10) using the information given on the variation form in Figure 2.9 on page 19 (addition of two further 13 A socket outlets to the reception area of Rachel Louise Hairdressing). This job will occupy about 6 hours.

SMITH & SONS Electrical Contractors

DAYWORK SHEET

Customer:

Job No.:

Date	No. of men	Time Started	Time Finished	Total for day	Travelling Time	Mileage and fares	Notes
	TOTALS						

Materials

Quantity	Catalogue No.	Description	For Office Use

Foreman's Signature   Date

Customer's Signature

Figure 2.10

The paperwork

At some stage in the preparation for starting an installation a list of materials has to be prepared. On large jobs this may have been carried out when the installation was designed but on smaller contracts it may be the responsibility of the electrician. For a number of reasons it may not be possible to visit the site before work is to start. This means that material lists have to be taken from drawings.

If it is a small installation the electrician may have all the materials that are required right from the time the job starts. Many contractors have their materials delivered to the site by the wholesaler. In such cases it is important to check that what is delivered is what was ordered and to check there is no damage to the goods.

Where extra materials are required on site a requisition is generally made out (Figure 2.11). An order is then written listing the requirements and it is sent to the wholesaler requesting delivery to the site (Figure 2.12).

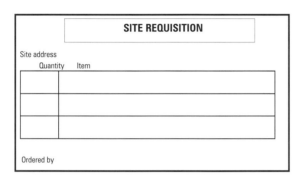

Figure 2.11   Site requisition form.

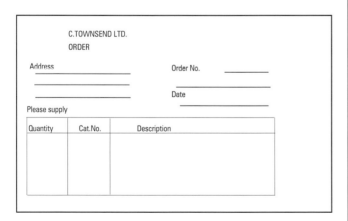

Figure 2.12   Order form.

The wholesaler will usually require a signature on a delivery note as proof that the materials and equipment have been received (Figure 2.13). Care must always be taken to ensure that what is on that delivery note is actually received. Trying to prove that something was not delivered which has been signed for can be very difficult.

Figure 2.13   Delivery note.

An invoice will be sent to the contractor for payment (Figure 2.14). Invoices are often sent out on a monthly basis and checks have to be made, with site representatives, to ensure that all of the materials on the invoice have been received.

Figure 2.14   An invoice.

Try this

Using a supplier's catalogue complete the order form in Figure 2.15 by filling in appropriate catalogue stock numbers.

C.TOWNSEND LTD.
ORDER

Address _____   Order No. _____

Date _____

Please supply

Quantity	Cat.No.	Description
3		150 mm standard cable tray
5		20 mm white conduit terminal boxes
2		pendant sets with 9 in cable
1		cooker control switch

Figure 2.15

What kind of information do you need to fill in on the following types of form:

A time sheet

A job sheet

A day work sheet

A variation order

Material lists are sent to the wholesaler on an
_____ form.

When the materials arrive on site a signature will be required on a _____ confirming that all materials ordered have been received.

Before a company pays an invoice checks will have to be made. What checks are these?

Part 3

On site

Relationships on site

When you work on a large site you may find yourself working with a **contractor**, other **subcontractors** and **clients** (Figure 2.16). On a large contract the client may have a broad outline of the plans and specifications drawn up by an **architect and consultant**.

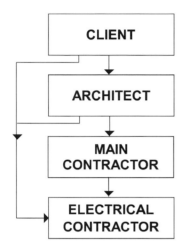

Figure 2.16

The client

The client is the person having the work carried out. On a small domestic job the client is likely to go directly to the electrical contracting firm. When a large contract is involved the client may go to an architectural practice or consultants.

The architect

The architect or consultant will produce plans and specifications to cover such items as the equipment to be used and where it will be located in the building. When the planning is complete the main contractors will be invited to tender for the job.

Main contractor

The firm that is successful will then engage subcontractors to carry out the specialist work, such as the electrical installation.

The electrical contractor

The electrical contractor, whether it is for a small domestic job or a major wiring contract, will be carrying out the wishes of the client.

On-site supervision

The site foreman will:
- supervise the work of the electrical installation team
- ensure that the work is progressing to schedule
- ensure that health and safety regulations are being adhered to
- act as the company representative
- keep a site diary
- produce "as-fitted" drawings

In a site diary notes are kept regarding the progress of work, hours worked by the electricians on site, details of site meetings, variation orders, accidents and any incidents that may have occurred such as damage to materials.

Good industrial relationships are important to the smooth running of the job. If there is a dispute between contractors and subcontractors or between subcontractors it can mean that the job is delayed, and this is likely to prove costly to those responsible for the delay.

The contract

In the world of business a contract is usually in the form of a written legal agreement. This agreement will cover all aspects of the terms and conditions of the job. Both the contractor and the client involved in the contract must agree to all the terms and conditions before the contract is signed.

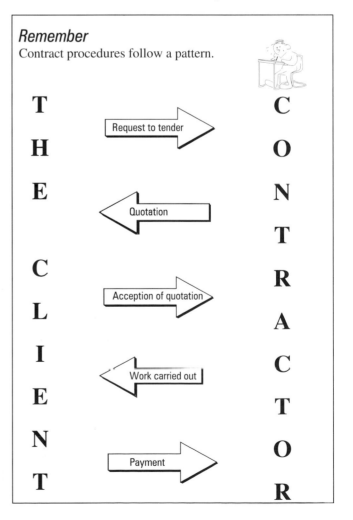

Remember
Contract procedures follow a pattern.

A contract would need to include, at least, such factors as:
- where the work is to be carried out
- when it will be started and when it is to be finished
- exactly what is included
- what standards it will conform to and
- which wiring system will be used

If a specification was issued then the contract will have been tendered to this specification. The electrician must ensure that the work is carried out to the specification.

The client

You may find that some clients/customers will be aware of the technicalities of the work while others will have no knowledge at all... some will be cooperative while others may be difficult and obstructive.

If you have a client that appears to have little or no understanding of what the job entails it is important not to make him feel foolish. Explain simply and clearly any parts of the work that are not understood. It may be necessary to ask questions of the customer in order to find the extent of their understanding.

A courteous approach may result in the customer returning with more work in the future. Customers who are not treated with respect will be likely to complain and they may also become aggressive.

Your employer will expect all staff on site to be representatives of the company whether dealing with clients, the contractor or subcontractors.

Site visitors

On occasions it is necessary for visitors such as inspectors and company's or customer's representatives to call at the work location. For example, representatives of all the trades working on a site will normally attend site meetings. Usually a system is set up to establish the identity of and the authority of the visitor and the purpose of their visit.

Visitors will normally go directly to the site office and will then be asked to provide proof of their identity and they will be required to sign in. (A record is kept of who is on the site at all times.)

Visitors must be informed of any safety equipment which they will require on site and any Health and Safety rules which apply. Any information given to the visitors must be clear and easily understood. They may need to be accompanied around the site, in which case it may be necessary to refer them to an appropriate escort.

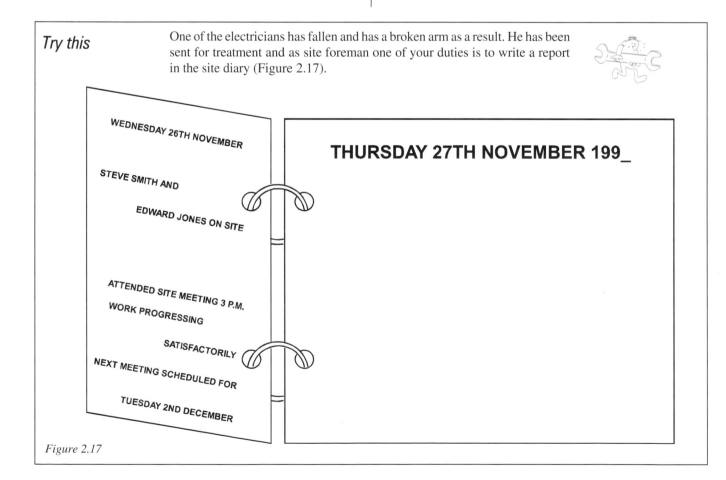

Try this

One of the electricians has fallen and has a broken arm as a result. He has been sent for treatment and as site foreman one of your duties is to write a report in the site diary (Figure 2.17).

WEDNESDAY 26TH NOVEMBER

STEVE SMITH AND

EDWARD JONES ON SITE

ATTENDED SITE MEETING 3 P.M.

WORK PROGRESSING

SATISFACTORILY

NEXT MEETING SCHEDULED FOR

TUESDAY 2ND DECEMBER

THURSDAY 27TH NOVEMBER 199_

Figure 2.17

Figure 2.18   I won't need a hard hat!

Points to remember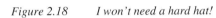

Whether you are in the office with your colleagues, representing your company on site, working with other trades or with clients, you should be courteous, polite, professional, and work to achieve and maintain good relationships. First impressions are important and this applies to both manner and dress.

It is important to understand the structure of your company and the role of each member of the team. Records and reports should be completed in the correct manner for your place of work.

A contract is a legal agreement between a contractor and client and either party may be sued for failure to comply with the terms.

Write in your own words what you feel the role of the following personnel is likely to be:

The client

The architect

The main contractor

The electrical contractor

The site foreman

Note down some of the items that a contract will contain:

What information should be given to every visitor to a site?

Try this

Assume that you are responsible for a visitor to the site on which you are working. Explain in your own words the procedures that you would follow.

Self-assessment multi-choice questions

Circle the correct answers in the grid below.

1. The most appropriate staff in an electrical contracting company to draw up a specification would be
   (a) clerical staff
   (b) drawing office staff
   (c) personnel staff
   (d) design staff

2. The most appropriate staff in an electrical contracting company to price a job would be
   (a) stores staff
   (b) electricians
   (c) estimating staff
   (d) design staff

3. Which of the following would be the most appropriate for the electrician to fill in on a daily basis when extra work has to be charged for?
   (a) a day work sheet
   (b) a time sheet
   (c) a job sheet
   (d) a variation order

4. Site diaries are usually kept by
   (a) the client
   (b) electrical site foreman
   (c) architect
   (d) main contractor's secretary

5. (i) Visitors to a site should be told of safety requirements.
   (ii) Visitors to a site should identify by name the person they wish to see.
   (a) Only statement (i) is correct.
   (b) Only statement (ii) is correct.
   (c) Both statements are correct.
   (d) Neither statement is correct.

Answer grid

1	a	b	c	d
2	a	b	c	d
3	a	b	c	d
4	a	b	c	d
5	a	b	c	d

26

3

Keeping Safe at Work

Complete the following to remind yourself of some important facts from the previous chapter.

How can you and your company make good first impressions on future customers?

(a) when they visit the company's premises in person

(b) when they telephone for some information

What information is contained on the following records?

time sheet

day work sheet

variation order

site requisition

On completion of this chapter you should be able to:

◆ explain employers' and employees' responsibilities under the Health and Safety at Work Etc. Act
◆ recognise the need to work with others to create a safe working environment
◆ describe the procedure for reporting accidents
◆ describe methods of circuit isolation
◆ describe the procedure to follow in the event of somebody receiving an electric shock
◆ recognise the need to inspect electrical equipment for faults before using it
◆ describe methods of fire prevention and recognise the fire-fighting equipment for different types of fire
◆ complete the revision exercise at the beginning of the following chapter

Part 1

Health and safety

You could find it useful to look in a library for copies of the regulations mentioned in this chapter. Read the appropriate parts and be on the look out for any amendments or updates to them.

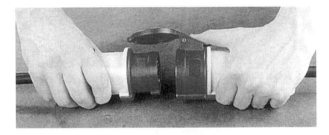

Figure 3.1   *BSEN 60309 Plug and socket.*

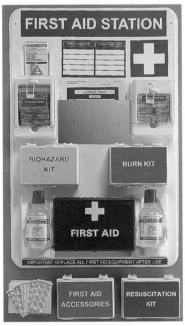

Figure 3.2   *First aid station.*

Keeping everyone safe at work is the responsibility of the employer *and* the employee (Figure 3.3), and both are required, by law, to observe safe working practices. Various Acts of Parliament govern what employers provide in a workplace and how the employees use this provision.

Health and Safety at Work Etc. Act 1974

This Act applies to everyone who is at work.

It sets out what is required of both
- employers and
- employees.

IS THIS WHAT IS MEANT BY EMPLOYER AND EMPLOYEE WORKING SAFELY TOGETHER?

Figure 3.3   Are they keeping safe?

The aim of this Act is to improve or maintain the standards of Health, Safety and Welfare of all those at work.

A number of regulations and codes of practice have been introduced under the Health & Safety at Work etc. Act, including:
- Management of Health & Safety at Work Regulations 1992
- The Electricity at Work Regulations 1989
- Manual Handling Operations Regulations 1992

Other laws and regulations which deal with Health, Safety and Welfare at Work include:
- The Factories Act 1961
- Safety Representatives and Safety Committees 1977
- Notification of Accidents and General Occurrences Regulations 1980

Remember
Watch out for new laws regarding Health and Safety and for amendments to the existing laws.

The employer's responsibility

Employers are required to provide and maintain a working environment for their employees which is, as far as practicable, safe and without risk to health. The "working environment" applies to all areas to which employees have access. For example, corridors, staircases and fire exits are included, as are gangways (Figure 3.4) and steps.

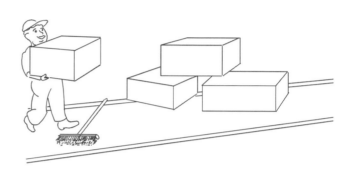

Figure 3.4   Keep gangways clear

The "safe working environment" includes such factors as maintaining a reasonable working temperature and humidity, adequate ventilation and fume and dust control. All areas must also be suitably and adequately illuminated.

Other facilities that are required by law include those for washing, sanitation and the supply of first aid equipment.

Employers are required to provide and maintain suitable safe tools and equipment for use by their employees and any training in the use of such equipment that is necessary Any information or supervision as may be required is also the employer's responsibility.

The employer must also ensure that the method of working is safe. Protective equipment, for example machinery guards, safety screens and protective clothing, must be supplied where required.

The storage, handling and transporting of goods is also the responsibility of the employer. Goods should be stacked on suitable shelves and in a suitable manner that will prevent danger. Some materials, especially chemicals, should be stored in the correct containers and labelled clearly. Heavy items may need to have mechanical handling aids, such as a fork-lift truck (Figure 3.5), for their safe transportation.

Figure 3.5   *A fork-lift truck.*

Figure 3.6   *A safety policy must be available to all employees.*

Try this

Which of the following are required by law to be provided by the employer?

	Yes	No
Hand washing facilities		
First aid equipment		
Kettle		
Free laundry		
Protective eye wear (where required)		
Tea/coffee facilities		

Employers must ensure that in the event of an accident, however slight, the details are recorded in a register kept at the workplace. The employer may be required to report these accidents to the Health and Safety Executive who may wish to investigate further. They may also inspect the accident register at the same time.

Major injuries that result in fractured bones, loss of sight or hospitalisation for more than 24 hours must be recorded in the accident register. In addition, every accident involving an employee must be notified to the Health and Safety Executive if the employee is unable to work for three or more consecutive days.

It is also a legal requirement for an employer with five employees or more to produce a safety policy and to make it available to all employees (Figure 3.6). This is usually in the form of a written notice, but it could be stored on a computer; however, it must be available for use. The safety policy will include details of any hazards and risks present in the working environment and the safety procedures which need to be taken in order to protect the health and safety of all persons concerned. This safety policy should be subject to regular review in consultation with safety representatives.

Site visitors

Employers are responsible for ensuring that visitors to the site are included when drawing up their safety policy. You will remember that authorised site visitors should be asked to identify themselves and identify by name the person they wish to see. Then they should sign in and be made aware of the safety procedures they need to follow while they are on site.

Try this

Take a look at your company's safety policy and make a note of the main points below.

The inspectorate

Employers have many legal duties with regard to Health and Safety at Work, and some of the main responsibilities have already been listed. Breaking these laws can result in prosecution.

The inspectorate are the Health and Safety Executive and they are concerned with the safety and welfare of everybody at work. Whilst they have legal powers to prosecute offenders of the law, it is far more likely that they will issue either an improvement notice or a prohibition notice. Only if these are not acted upon will prosecution follow.

Improvement notice

An improvement notice can be served on an employer if an HSE inspector is satisfied that a statutory regulation has been contravened.

Once this notice has been served the employer has a specified time (not less than 21 days) in which to appeal to an Industrial Tribunal. During the time that the appeal is pending, the improvement notice is suspended and the equipment may be used. If the employer fails to comply with an improvement notice within the specified time then prosecution will almost certainly follow.

Prohibition notice

A prohibition notice can have the effect of stopping an activity or practice immediately, without recourse to appeal, if the inspector is satisfied that there is a risk of personal injury. Failure to comply with this notice will almost certainly result in prosecution.

Remember

An employer must provide:

- a safe place of work with safe access and exit
- safe equipment and system of work
- a safe working environment including facilities for sanitation, washing and first aid
- safe methods of handling and storing goods
- accident procedure and register
- training and supervision
- a safety policy
  (for workplaces with 5 employees and over)

Points to remember ◀ – – – – – – – – – – – – –

Who does the Health and Safety at Work Etc. Act cover?

_____

What is required of them?

_____

The aim of this Act is to maintain or improve

_____.

The employer must provide and maintain a safe working environment. What areas in your place of work are covered by "the safe working environment"?

Complete the following list of factors that must be included in "the safe working environment".

> reasonable working temperature
> reasonable working humidity

Employers must also provide and maintain safe equipment and

_____

On occasions you may be required to wear

> eye protection
> hearing protection
> head protection
> hand protection
> foot protection
> a safety harness or belt

An employee has an accident and is unable to work for four consecutive days. To whom must the accident be reported and where should it recorded?

If your company has machinery that fails to comply with legal requirements then The Health and Safety Executive may issue either of the following two notices:

Part 2

Preventing accidents

Employees' responsibilities

An *employee* can be prosecuted for breaking safety laws.

Employees are required by law to:
- take reasonable care for their own health and safety and not to endanger others
- cooperate with their employer on health and safety procedures
- not interfere with tools, equipment etc. provided for their health, safety and welfare
- correctly use all work items provided in accordance with instructions and training given to them

Accidents at work
- Accidents are **unplanned**.
- Accidents do not **"just happen"**.
- Accidents are **caused by people**.

An *accident* is an unexpected or unintentional event which is often harmful.

There are three main causes:
- people who become unsafe because of such factors as boredom, horseplay, carelessness or lack of knowledge
- people who have provided or maintained an unsafe environment
- people who have misused a safe one.

Accidents can be prevented by observing the following:

Make the environment as safe as possible.

Keep the floors and workbenches clear of rubbish, putting any tools and equipment away when they are not needed. Replace caps on bottles (Figure 3.7) and if any liquids are spilled clean them up immediately. Waste materials should be removed or stored in suitable containers. Access routes must be kept clear at all times and temporary hazards, such as where a floorboard has been removed, should be adequately guarded.

Figure 3.7   *Replace caps on bottles.*

Do not interfere with machinery. It is extremely dangerous to remove guards. A machine must not be used by someone who has no authority to use it or who has not been suitably trained. Long hair should be tied back and loose clothing should not be worn as it may get caught in machinery.

Use the correct tool for the job and make sure it is in a safe condition. Tools which show signs of damage or wear can be potentially dangerous and should not be used until repaired or replaced (Figure 3.8).

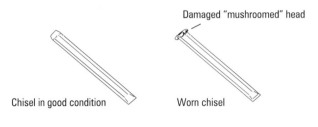

Damaged "mushroomed" head

Chisel in good condition   Worn chisel

Figure 3.8

Wear protective clothing and other equipment if it is considered necessary. For example, eye protection (Figure 3.9) should be worn when grinding metals and hard hats should be worn on all building sites. Protective clothing and equipment should conform, where necessary, to the mandatory standards.

Figure 3.9   *Wear eye protection.*

Read printed rules, warning notices and instructional posters, which should be prominently displayed. Other information, instruction and training may be given which is equally important and must be followed.

Safety signs give warning of particular dangers and will show the particular type of protection required for the conditions concerned.

In your place of work find out the 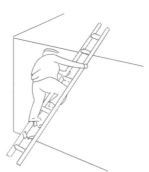 following:

1. Who do you have to report damaged tools to?

2. Who issues protective clothing and equipment?

3. The first aid box at your place of work, on site, at your training centre or using the photograph at the beginning of this chapter contains:

Would you know how to use the items you have listed?

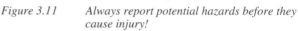

Figure 3.11 *Always report potential hazards before they cause injury!*

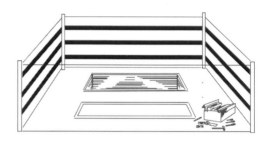

Figure 3.12 *For safety always place a guard around a drop.*

Employees should also take reasonable care of their own health and safety and that of others (Figure 3.10). This means that they will be expected to do their job in a sensible way and they should not run around or fool about. They should also make sure that they are fit for work and not overtired or ill or suffering from the effects of drink or drugs.

Always report an accident whether it results in injury or not as it may prevent another more serious accident from the same cause.

It is important to note where the first aid box is located; this is identified by a white cross on a green background (Figure 3.13). Find out who are the appointed first-aiders.

Figure 3.13 *First aid – white cross on a green background.*

Figure 3.10 *Work sensibly.*

Employees should look out for hazards in the workplace and report them (Figures 3.11 and 3.12). You may be the first to notice, and if the hazard is ignored it may cause an injury to you or your colleagues. This will include deficiencies in equipment, such as broken ladder rungs, and insecure structures, falling objects and so on.

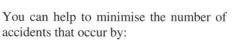

Remember
Accidents don't just happen – they are caused!

You can help to minimise the number of accidents that occur by:

- learning the safety rules of your workplace and industry and following them
- protecting yourself and the people around you by using the correct equipment, clothing and gear
- keeping a look-out for potential hazards
- behaving sensibly and thoughtfully.

Try this

Find out where the accident book and the first aid box are kept at your place of work.

The accident book is located

The first aid box is located

The appointed first-aiders are

Discuss with your friends and colleagues, at work or at your training centre, any potential hazards which any of you have noticed. Note down any which have not been referred to in this chapter and also what action was taken.

Following any accident employees need to note the following details in order to fill in the accident record which must be kept in any workplace:

- the date of the accident or dangerous occurrence
- the place where the accident or dangerous occurrence took place
- a brief description of the circumstances
- the name of any injured persons
- the sex of the injured persons
- the age of the injured persons
- their occupations
- the nature of the injuries.

There may be other details required by particular organisations. These may range from witnesses to suggestions for the prevention of further accidents. Major injuries will have to be reported to the appropriate authority.

There are other guides available from the HSE, industry and workplaces that will help in understanding the employees' responsibilities.

Just as employers are responsible for providing training, information and supervision, so the employees are responsible for carrying out the work in the manner in which they have been trained. Employees are also responsible to the customer in following instructions from their employer, including handing over equipment manufacturers' instructions and removing waste materials from the work site.

Points to remember ◄ – – – – – – – – – – – – – –

These points are important, learn them off by heart.

To prevent an accident:
- keep the work area clean and tidy – clear up the rubbish!
- take care near machines – don't interfere with safety guards and be careful not to have long loose hair or clothing
- take care of your tools – keep them in good condition
- wear protective clothing or equipment when it is necessary
- take note of any rules, regulations or safety signs and obey them
- take reasonable care for your own health and safety and do not endanger others
- look out for hazards and report them – do not assume that this has already been done
- always report an accident and note down the details required to fill in the accident form
- know where the first aid box is located and who is the appointed first-aider

Part 3

Electrical safety

Electrical accidents (or accidents involving the use of electricity) may result in electric shock, fire and burns.

Accidents can occur when materials or equipment are faulty, poorly maintained or misused. Misbehaviour or carelessness, and protective devices and equipment which are not used or are used incorrectly are also contributing factors.

All these factors apply to electrical accidents.

In order to reduce the possibility of accidents when using electricity, strict observance of the applicable laws, standards and codes of practice is imperative.

Laws which must be observed include:

The Health and Safety at Work etc. Act 1974,

The Electrical Supply Regulations 1988 and

The Electricity at Work Regulations 1989

BS 7671, the Regulations published by the Institution of Electrical Engineers (known as the IEE Wiring Regulations) are not law but are accepted as standard practice in electrical installation work.

The **British Standards Institution (BSI)** also produce standards and codes of practice. Those applicable to electrical installation work should be complied with.

These laws, standards and codes of practice relate to the manufacture, installation and use of electrical equipment. Before working on electrical equipment all staff should be properly qualified, trained and competent in their work.

There are some general safety rules to follow. There may be others that particular organisations or industries have and it is important to know which rules apply and ensure they are followed.

General safety when using hand-held portable electrical equipment

When portable electrical equipment is used in the workplace or on construction sites, accidents can often be prevented by following a few simple rules.

A visual check (Figure 3.14) on cables and plugs, which are particularly liable to damage, can prevent a serious accident.

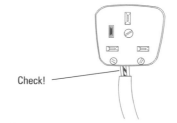

Figure 3.14   Visually check cables and plugs.

Electrical equipment may develop faults which do not affect its operation. The equipment may, however, present a potential hazard. Testing the equipment and implementing repairs help to ensure accidents are prevented (Figure 3.15).

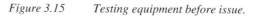

Figure 3.15   Testing equipment before issue.

In general all electrical equipment should be connected to earth through a circuit protective conductor. However, if the equipment conforms to the appropriate standards it may be classed as Class II equipment often called double insulated. The classes of equipment are described in BS 2754:1976 "Construction of electrical equipment for protection against electric shock", which is is based on IEC Report 536. The symbol in Figure 3.16 would be displayed on the case of the equipment and in these situations no circuit protective conductor would be necessary.

Figure 3.16   British Standard symbol for Class II (double insulated) equipment.

Where equipment is used out of doors or in damp environments it should have a residual current device (RCD) protecting the circuit with an operating current of 30 mA (Figure 3.17). This means that should a fault develop between phase and earth or neutral and earth the circuit would automatically switch off before the fault current could reach 30 mA.

Figure 3.17   RCD.

On construction sites or in factories it is advisable to limit the voltage used on hand held portable equipment to 110 V a.c. This is supplied through a transformer, as shown in Figure 3.18, which limits the line voltage to 55 V above earth potential. This is generally regarded as the maximum safe working voltage for a.c. On d.c. the maximum safe working voltage is 120 V.

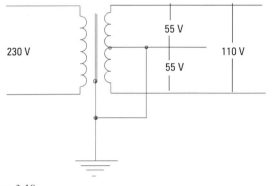

Figure 3.18

Restrictive conductive locations

Where portable equipment is to be used in conductive locations where physical movement is restricted, such as tunnels and boilers, BS 7671 identifies particular requirements for these conditions.

Although in domestic premises the 13 A plug is used to connect portable equipment to the supply, in industrial locations a plug conforming to BSEN 60309-2:1992 (previously known as BS 4343) is used (Figure 3.19). There are different colours of plug, socket, connector and appliance inlet in order to identify the voltages they are designed to be used on. Hence a blue plug can only be connected to a blue socket. So that a plug cannot be connected to the incorrect voltage an interlocking lug is located at different positions, for different voltages, as illustrated in (Figures 3.20 and 3.21).

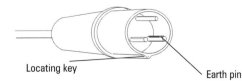

Figure 3.19   Plug conforming to BSEN 60309.

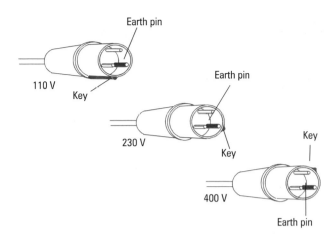

Figure 3.20   BS 60309 plugs of different voltages.

Table 3.1

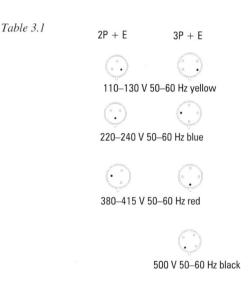

Figure 3.21   Examples of the position of the earth relative to the interlocking lug for different voltages.

Colours of plugs	
Voltages	**Colour**
50 V	white
110–130 V	yellow
220–240 V	blue
380–415 V	red
500 V (50–60 Hz)	black
> 50 V (100–300 Hz inclusive)	green
> 50 V (300–500 Hz)	green

Isolating the supply before work

The normal practice required under the Electricity at Work Regulations is to work with the equipment *dead* (Figure 3.22). A "dead" circuit is one that is neither "live" or "charged". This also means that a method of preventing accidental reconnection of the supply should be employed to protect the person carrying out the work.

DANGER!

*Figure 3.22   It is **not** normal practice to work on live equipment.*

So suitable means must be found to cut off the supply of electrical energy and then to isolate the electrical equipment, which when "isolated" cannot be inadvertently reconnected.

The supply can be cut off by removing a plug or a fuse, but once this is done steps must be taken to ensure that they are not replaced. When the supply has been switched off, to make sure it is not accidentally reconnected it must be *locked off*. This incorporates some form of locking device such as a padlock, and one key which is kept by the electrician doing the work (Figure 3.23). Only when this has been done is it said to have been *isolated*. The means of isolating equipment should be clearly labelled so that there is no doubt about which equipment it controls.

THAT'S
BETTER!

Do not switch on Men working on equipment

Figure 3.23

Caution notices should be posted where equipment is isolated warning that work is being carried out on the equipment and that it would be dangerous to reconnect it. If there is any *live* equipment adjacent to isolated equipment then danger notices should be posted. These notices must be removed when they are no longer required.

Before work can be started the conductors must be proved "dead". Test instruments (Figure 3.24) or voltage indicators are used for this and they must be checked both before and after testing to ensure that they are working correctly.

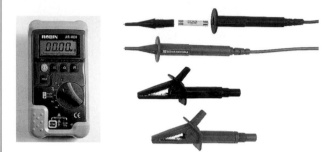

Figure 3.24   Test instruments.

A more sophisticated form of protection is the *permit-to-work*. This is a system which provides a written job description for the worker to follow, lists the safety precautions which should be taken and finally provides the operator with a key and padlock for locking off the isolator. It should also provide any warning notices which may be necessary. A permit-to-work should never be modified.

The Electricity at Work Regulations state that where work must be undertaken on "live" circuits strict safety procedures must be followed.

The requirement of these safety procedures are:
- there is absolutely no way of doing the work with the circuit dead
- the person doing the work has been fully trained and is competent to carry out the work
- appropriate instruments, tools, equipment and protective clothing are used
- they have been authorised to do the job
- they place warning notices and, where appropriate, erect barriers and enclosures to protect anyone else who may be in the vicinity

Figure 3.25   Heavy duty lockout tags.

Points to remember ◄ – – – – – – – – – – – – –

Electricity when used correctly can be a great help to everybody, but if it is not used safely it can be a killer. Some electrical equipment is made to a standard of insulation so that an earth connection is not required.

Generally if a metal case is used on electrical equipment it must be connected to earth using a _____.

When portable electrical equipment is used on construction sites or in factories _____ voltages are usually used for safety.

What do the letters RCD stand for?

Explain in your own words what an RCD does.

In industrial locations a plug is used conforming to BSEN _____.

What ensures that the plug cannot be connected to the incorrect voltage?

Try this

The electric drill you have been allocated for your work has a frayed cable.

What should you do?

List four likely faults which may occur on domestic equipment and what steps would need to be taken in each case to prevent an accident.

Part 4

Shock, fire and burns

Electric shock

If a person comes into contact with a phase conductor while also in contact with earth, current will pass through the person's body (Figure 3.26). It can also happen if a person touches two live conductors as the body will complete the circuit between them. In either case the person concerned will have received an electric shock.

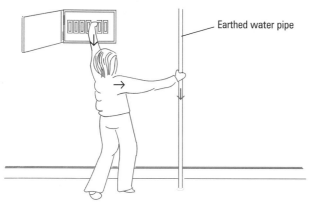

Earthed water pipe

Figure 3.26

Electric shock may be caused by the metal case of an electrical appliance becoming "live". This may occur if, for example, the circuit protective conductor on an appliance is broken or not connected and the phase conductor makes contact with the metal casing, which in turn becomes "live". Then anyone touching the casing and other metalwork could become "a conductor" and suffer an electric shock.

If you find someone who has received an electric shock your first priority must be to take care not to become a casualty yourself.

For this reason there is a correct procedure that should be followed when dealing with a person who has received an electric shock.

FIRST the connection to the phase conductor must be broken without causing any further injury:

> **IF** possible, cut off the electricity supply

> **IF** this is not possible the casualty will have to be pulled clear **BUT** only with a dry insulator ("dry" because water conducts electricity). A dry insulator could be rubber gloves, a newspaper, a rope or a wooden pole. Alternatively, standing on a rubber or plastic mat or dry wood could prevent you receiving an electric shock. Do **NOT** touch the casualty's bare skin without taking precautions to prevent electric shock.

Once disconnected from the supply

> **IF** the casualty is not breathing start resuscitation immediately (Figure 3.27) and call for expert help.

> **IF** the casualty is unconscious, but breathing, place them in the recovery position (Figure 3.28) and get help. The injured person may be burned or bleeding as a result of the shock or fall and require further medical aid.

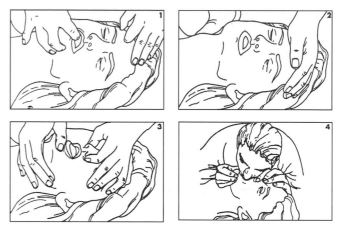

Figure 3.27   *Resuscitation.*

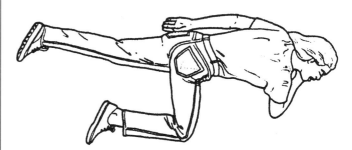

Figure 3.28   *The recovery position.*

Electrical burns

High currents can create arcs that may cause serious burns. With high current the voltage may be very low, so the electric shock potential is slight.

A good example of how a high-current, low-voltage electrical burn can be caused is the misuse of a lead acid battery as used in a car. The voltage is only 12 V d.c., but if the battery is shorted out the current can be over 100 A. A spanner carelessly laid down or dropped across the terminals of such a battery will cause an arc. This produces considerable heat and may cause very serious burns.

If you or your colleagues receive an electrical burn get expert medical help immediately.

Fire hazards

In order for a fire to start there must be fuel, air (oxygen) and heat (Figure 3.29). If all three are not available then a fire will not start. If a fire has started, then by removing any one of the three the fire will be extinguished.

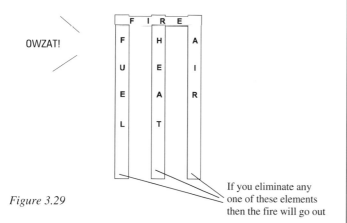

OWZAT!

If you eliminate any one of these elements then the fire will go out.

Figure 3.29

For example the use of a foam fire extinguisher cuts off the supply of oxygen. Alternatively, if there is no more fuel, combustible material, then a fire will go out.

Enough heat can be produced during initial combustion to maintain or accelerate the fire, resulting in rapid spreading of the fire.

Fires can be started in numerous ways such as people smoking carelessly (Figure 3.30), friction heat, sparks, naked flames, fuel leaks and faults or failures in equipment resulting in overheating and chemical reactions.

Preventing fires

There are precautions that we can take to prevent fires from starting, many of which are common sense. We should take care to store combustible materials away from heat sources, and all areas should be kept free from combustible rubbish and dust.

Equipment which could cause a fire should be regularly maintained and serviced, including electrical equipment and associated cables, plugs and flexes. Fuel pipes should be checked for leaks and flammable materials should be suitably stored at controlled temperatures.

DANGER
NO SMOKING

Figure 3.30   Enforce the no smoking rule in areas where it could start a fire.

If a fire occurs...

Make sure that you understand the hazards involved and know what to do in the event of a fire.

Places of work are required to have a fire procedure and if a fire occurs these should be followed immediately.

This will generally involve

- raising the alarm
- calling the fire service
- using the correct fire fighting equipment for the particular type of fire where appropriate
- shutting down equipment if possible
- clearing personnel from the area to the fire assembly point
- shutting all doors when evacuating the premises to prevent the spread of fire and smoke and to limit the extent of damage to property
- reporting to the supervisor at the prearranged assembly point

The fire drill procedure that is displayed in your place of work must always be followed. Make sure you are aware of the emergency procedures which cover evacuation in the event of other emergencies, for example in the case of an explosion, contaminated fumes or terrorist activity.

Regulations covering fire safety include

- Health and Safety at Work etc. Act 1974
- Fire Precautions Act 1971 and Regulations
- Building Regulations 1976
- Fire Services Act 1947.

Try this

Find out the fire drill procedure either at your place of work or your college/training centre.

Make a note of the procedure below.

Fire safety equipment

It is important to find out the location of all available fire fighting equipment and the types of fire for which it can be used.

Fire extinguishers

All British-made fire extinguishers are colour-coded to indicate their particular purposes, Table 3.2 shows the different types and main uses. From 1 January 1997, all new certified fire extinguishers, under BS EN 3, must have red bodies. In the UK colour identification panels are placed on or above the operating instructions. Fire extinguishers with full body colour codings may still be found and can continue to be used until they need replacing. Halon extinguishers, colour code green, are being phased out of service due to the adverse effects they have on the environment.

Fire blankets

Fire-resistant blankets are used to cut off the air supply to extinguish the fire. These are generally used on fat or oil fires, for example when a deep fat fryer catches fire, and they can also be used to smother clothing fires.

Burn injuries

If you are to deal with minor burns you should first wash your hands and then flush the burn with plenty of clean, cool water, do not remove clothing that may be sticking to the burn and apply a sterilised dressing. If the burn is serious, get expert medical help immediately.

Table 3.2

Type of fire	Water (red panel)	Foam (beige panel)	Dry powder (blue panel)		CO$_2$ gas (black)
			Dry powder to BS 5423	ABC Dry powder to BSEN 3 1996	
Class A Paper, wood, textiles	✓	✓		✓	
Class B Flammable liquids such as oil, paint, petrol, paraffin, grease		✓	✓	✓	✓
Class C Flammable gases such as LPG, butane, propane, methane			✓	✓	✓
Electrical hazards			✓	✓	✓

Emergencies

An emergency is any event which requires immediate action and in the event of an emergency your priority is to limit injury to persons before limiting damage to property.

You should:

- raise the alarm
- notify the professional emergency services
- suspend work immediately
- isolate equipment from its power source if it is safe to do so
- then proceed in accordance with safety procedures to a recognised assembly point.

Points to remember ◀ – – – – – – – – – – – –

It is important to know how to treat a person who has received an electric shock, without causing further accidents and injury.

It is better to prevent a fire occurring if possible, but should fire break out then the correct extinguisher must be used.

Keeping safe means that you must:

- stay alert
- be aware of how to protect yourself and others
- know what to do in an emergency
- report all hazards you cannot cope with yourself

If you find someone who has had an electric shock, what must your priority be?

What three factors have to be present for a fire to occur?

Give three examples of how fires can be started.

Which fire extinguisher should not be used on a fire of flammable gases?

Which colour code of fire extinguisher can you *only* use on a paper/wood/fabric fire?

Add to these lists any other items *you* are aware of.

Accidents may be caused by:

> human error
> irresponsible behaviour
> carelessness
> improper dress
> lack of training
> lack of supervision
> inexperience
> tiredness
> drug abuse
> alcohol abuse
> environmental conditions
> unguarded machinery
> faulty tools
> inadequate ventilation
> untidy workplaces
> poor lighting
> misuse of, or interference with, equipment
>   provided for health and safety

Accidents may be prevented by:

> eliminating hazards
> minimising hazards
> guarding hazards:
>   safety screens
>   barriers
>   fences
> using protective clothing
> working responsibly
> thinking ahead
> staying alert
> checking

Try this

See how many examples of fire-fighting equipment you can find.
Note the type of equipment, its location and the use to which it can be applied.

E.g. extinguisher with red label (water) caravan wood/fabric

_____

_____

_____

_____

_____

_____

Self-assessment multi-choice questions

Circle the correct answers in the grid below.

1. The Health and Safety at Work Etc. Act applies to
   - (a) employers only
   - (b) employees only
   - (c) both employers and employees
   - (d) safety officers only
2. Under Health and Safety Law which of the following is the responsibility of the employer to provide?
   - (a) tea making equipment
   - (b) free laundry service
   - (c) safety training
   - (d) weekly safety reports
3. An employer is required to report an accident to the Health and Safety Executive if an employee is
   - (a) unable to work for more than three days due to the accident
   - (b) at a hospital casualty department for a morning before returning to work
   - (c) treated at work for a cut hand
   - (d) sent home for 24 hours due to the accident
4. An Improvement Notice is issued by
   - (a) an employee to an employer
   - (b) a representative of the Health and Safety Executive
   - (c) a works safety officer
   - (d) an employer to an employee
5. A first aid box is identified by a
   - (a) white cross on a green background
   - (b) red cross on a white background
   - (c) white cross on a black background
   - (d) green cross on a white background
6. The maximum safe working voltage on a.c., under normal conditions, is accepted as being
   - (a) 230 V
   - (b) 110 V
   - (c) 55 V
   - (d) 12 V

7. When the British Standard symbol, shown in Figure 3.31, is used on equipment this means that it

Figure 3.31

   - (a) is designed for use on 110 V
   - (b) must be earthed
   - (c) is covered by 12 months guarantee
   - (d) conforms to double insulated standards
8. The first action to be taken when finding someone suffering from an electric shock is
   - (a) phone for an ambulance
   - (b) disconnect the person from any electrical supply
   - (c) turn them over on their back
   - (d) run for help
9. The three elements that all fires require are
   - (a) fuel, air and heat
   - (b) fuel, air and oxygen
   - (c) fuel, heat and paper
   - (d) heat, air and oxygen
10. Which of the following fire extinguishers is correct for the type of fire?

	Extinguisher	*Fires involving*
(a)	foam	electricity
(b)	water	flammable liquids
(c)	foam	propane
(d)	dry powder	LPG

Answer grid

1	a	b	c	d	6	a	b	c	d
2	a	b	c	d	7	a	b	c	d
3	a	b	c	d	8	a	b	c	d
4	a	b	c	d	9	a	b	c	d
5	a	b	c	d	10	a	b	c	d

4

The Working Environment

Before starting this chapter, remind yourself of some important facts from the previous chapter.

What is the preferred voltage limit for hand-held portable equipment on a construction site?

What is meant by "the operating current" of an RCD?

Why are there different colours and designs of plugs conforming to BSEN 60309-2:1992?

What are the three elements necessary for a fire?

What fire-extinguishing equipment should you use on a burning chip pan?

Have you, friends or work colleagues ever worked in an area considered as hazardous? If so, make a note of any that you find and the reasons they were hazardous.

The electrical industry requires work to be undertaken in a wide range of work locations, and the environments of each of the different work sites may introduce additional hazards.

Figure 4.1   Emergency shower/eye wash.

Figure 4.2   Danger.

- ◆ recognise the need to take particular environmental conditions into consideration before starting work
- ◆ recognise the need to take additional precautions in situations where chemicals are present
- ◆ identify possible hazards connected with working in cold, wet or noisy environments
- ◆ identify the four categories of safety signs
- ◆ recognise the need, for your own safety and that of others, to obey any signs that apply to you
- ◆ complete the revision exercise at the beginning of the next chapter

Figure 4.3   Fire blanket safety sign.

Part 1

Hazardous areas

Chemicals
Chemical hazards (e.g. Figure 4.4) are covered by the Control of Substances Hazardous to Health Regulations (COSHH). These regulations require that all chemicals that could affect Health and Safety must have a *data sheet* which lists the hazards posed by the substance, any other chemical it must not come into contact with and additional information such as handling precautions.

Figure 4.4   Chemicals are used in electro-plating baths.

Employers are required to assess work which is liable to expose any employee to hazardous solids, liquids, dusts (Figure 4.5), fumes, vapours, gases or micro-organisms. They must evaluate the risks to health and implement action to remove or reduce those risks.

Figure 4.5   Flour stores are dusty environments.

Many precautions are well known, like only using adhesives in well-ventilated areas and not storing chemicals in drink containers. Some hazards are not so apparent. For example, if you were drilling into concrete to make a conduit fixing then it is possible that the concrete dust, when breathed into your lungs, could in the long-term cause damage (Figure 4.6).

Figure 4.6   Concrete dust may cause lung damage over long term.

Some substances, known as *irritants* and *allergens*, can cause problems such as dermatitis. Even though you may be using such substances, for example epoxy resin or wet cement, without any apparent problems, it could be many years later that dermatitis appears.

Care should also be taken before working with electricity near gas cylinders. It is worth checking that the gas has not been inadvertently left on and that you cannot smell or hear any gas near the cylinder, indicating that there may be a leak.

What is important?

Think about the work before you start and make sure that you are taking the correct precautions.

Noise

Figure 4.7

Ear defenders (Figure 4.7) should be provided by your employer and worn when working in a noisy environment. They should comply with current BS specifications, fit properly and be comfortable. Loud noise can cause temporary partial loss of hearing even if you are only exposed to it for short and infrequent periods.

If you have to work in a very noisy location for several hours every day this partial loss may become permanent and gradually degrade even further.

The Noise at Work Regulations were introduced to counteract these problems.

These regulations require that where an employer has assessed the risks of injury to hearing by loud noise as too high, and it is not practicable to reduce the noise emissions by changing equipment design, all employees working in the hazardous area should be provided with adequate ear defenders.

Ear defenders should be worn at all times when working in a noisy environment and you should ask your employer to check the noise levels if you feel that ear protection may be necessary.

Some power tools may generate significant noise and the manufacturers' recommendations will normally state that ear protection must be used.

Cold

If work has to be undertaken outside in the winter, *coldness* can be a considerable hazard.

Many people tend to forget that, even though the thermometer may indicate that the temperature is 10 °C, if there is a 20 miles per hour wind the effective temperature to a person will be 0° Celsius. Similarly, a temperature of about minus 1° drops to about minus 9° Celsius in a 10 miles per hour wind, and this effect is commonly known as the "wind chill factor".

An early indication of the body becoming cold is when exposed skin becomes very white. This indicates that the blood supply has been cut off to help insulate the more vital internal organs.

It is important when working in cold conditions to wear the correct protective clothing (Figure 4.8). Not only nice warm coats, gloves and socks, but it should be remembered that the majority of body heat is lost through the head and so a hat may be essential as well! Protective clothing needs to be made from materials which resist winds and rain but are also light and flexible.

In particularly severe conditions it is also very important to make sure that you have a hot meal and drink for your lunch (Figure 4.9).

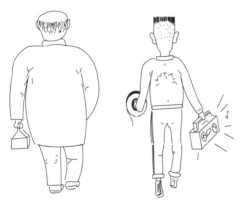

Figure 4.8   *Some people may find that a woolly hat, in addition to a warm coat, is essential protective clothing!*

Figure 4.9   *A hot drink/food for your lunch is very important when working in cold conditions.*

Wet and damp

Wet and damp locations (Figure 4.10) can amplify possible problems as the effects of coldness can set in much quicker. Using electricity can be particularly hazardous and adverse conditions can cause lack of concentration.

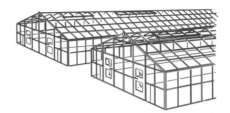

Figure 4.10   *Greenhouses may be damp.*

Where work is undertaken in adverse conditions additional care must be taken. This may include:
* working in shifts to reduce the time spent in the adverse conditions
* the use of extra low voltage tools and equipment to eliminate the risk of electric shock.

In all work situations and locations suitable clothing should be worn. This clothing should not introduce additional hazards – for example torn or loose clothing could get caught in equipment and flimsy footwear will not protect against hard and sharp objects and it could be advantageous on some jobs to wear overalls. If you need to wear eye or face protection check they conform to current BS specifications.

In cold or wet situations care must be taken, as surfaces may become slippery, particularly when carrying heavy or bulky objects which may create additional hazards if dropped. Likewise if you have cold or wet hands you may not be able to grip safely when holding tools or climbing ladders.

Try this

Your employer should have taken into account any hazards introduced by adverse working environments. Check any safety notices and policies and make a note of the hazards that have been identified and the precautions required to overcome or lessen the hazard.

Before starting work in any location take the environmental conditions into consideration.

Where chemicals are present additional precautions may need to be taken. Note any particular areas of your company's activities where such precautions have been required.

When working in a noisy environment you should wear ear protection or you may _____.

Working in cold, wet situations may cause the following hazards (add to this list):

    slippery surfaces
    lack of concentration

Try this

Your place of work has been asked to do some work in (a) a wine cellar, (b) a glue factory and (c) a paint store. What hazards can you think of that may require you to take precautions and/or use protective clothing in each case.

(a)

(b)

(c)

Part 2

Safety signs

"The Health and Safety (Safety Signs and Signals) Regulations 1996" states that all new Health and Safety Signs *must* contain pictorial symbols in addition to text. Other requirements contained in these regulations include:

- if a hazard cannot be adequately controlled by any other means then it must be marked with a safety sign
- every employer must ensure that their employees are trained in the meaning of safety signs
- any emergency escape route and fire fighting equipment must have their location identified with safety signs

Most workplaces have safety signs and posters to alert you of any dangers. These are grouped into four categories:

- prohibition signs
- warning signs
- mandatory signs
- safe condition signs

and in addition there are location signs for fire fighting equipment.

Prohibition signs

Prohibition signs are circular in shape and have a crossbar through the centre and mean STOP or DO NOT. They are red on a white background and must be obeyed.

For example (Figures 4.11–4.13):

Figure 4.11   No smoking.

Figure 4.12   No pedestrians.

Figure 4.13   Do not use.

Warning signs

Warning signs are triangular and yellow with a black border. They mean CAUTION, RISK OF DANGER or HAZARD AHEAD.

For example (Figures 4.14–4.16):

Figure 4.14   Danger – high voltage.

Figure 4.15   Corrosive.

Figure 4.16   Caution.

Mandatory signs

Mandatory signs are white on a blue background, circular and mean YOU MUST DO.

For example (Figures 4.17–4.19):

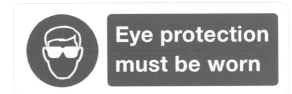

Figure 4.17   Wear eye protection.

Figure 4.18   Wear head protection.

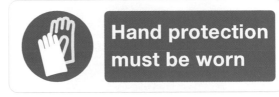

Figure 4.19   Wear hand protection.

Safe condition

Safe condition signs are rectangular and white on a green background and indicate the correct place to go or the correct action to be taken in an emergency. They give information about safe conditions.

For example (Figures 4.20–4.22):

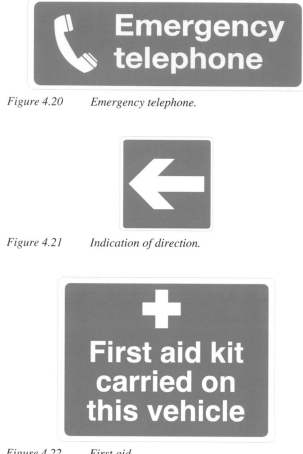

Figure 4.20   *Emergency telephone.*

Figure 4.21   *Indication of direction.*

Figure 4.22   *First aid.*

You are likely to find extra information included with any of the safety signs, such as the type of first aid available, the particular type of eye protection required, the clearance height of an obstacle etc.

For example (Figure 4.23):

Figure 4.23   *Emergency eye wash station.*

Try this

Design your own signs. Each sign must be the correct shape, and the correct colour must be clearly stated.

1. A mandatory sign which means "WEAR RESPIRATOR".

2. A warning sign which means "RISK OF EXPLOSION".

3. A prohibition sign which means "DO NOT EXTINGUISH WITH WATER"

4. An information (safe condition) sign which means "EMERGENCY SHOWER".

Fire fighting equipment signs

The Health and Safety (Safety Signs and Signals) Regulations require the location of all fire fighting equipment to be marked in red. Typical signs are given in Figures 4.24–4.26).

Figure 4.24   Fire extinguisher location sign.

Figure 4.25   Fire alarm call point.

Figure 4.26   Fireman's switch location sign.

Points to remember – – – – – – – – – – – – – –

You can help to keep yourself safe by:

- recognising the hazards presented by the environment you are working in.
- taking adequate precautions to protect yourself.
- following any safety procedures set out for you by your employer
- taking notice of safety signs and obeying them

There are _____ categories of safety sign. They are prohibition signs, _____ signs, mandatory signs and safe condition signs. Prohibition signs are _____ in shape and have a crossbar through the centre. They are _____ on a white background and must be obeyed.

Warning signs are triangular and _____ with a black border. They mean caution, risk of danger or hazard ahead.

Mandatory signs are white on a _____ background and _____ in shape. They mean you must do what the sign says.

Safe condition signs are rectangular and white on a _____ background and give information about _____.

Look around your place of work or your college for any of these categories of signs.

Self-assessment multi-choice questions

Circle the correct answers in the grid below.

1. The mandatory sign which means you must wear gloves shows
   - (a) green gloves on a white background
   - (b) blue gloves on a white background
   - (c) white gloves on a green background
   - (d) white gloves on a blue background
2. Warning signs are
   - (a) circular with a crossbar through the centre
   - (b) triangular with a black border
   - (c) rectangular with a green border
   - (d) circular with a blue border
3. Chemical hazards are covered by Regulations called
   - (a) COSHH
   - (b) BS7671
   - (c) EAWR
   - (d) HASAW
4. A prohibition sign
   - (a) shows what **must** be done
   - (b) warns of **danger**
   - (c) gives information of **safety** provision
   - (d) shows what must **not** be done
5. The sign which indicates the position of an emergency telephone is
   - (a) a white telephone on a blue background
   - (b) a white telephone on a yellow background
   - (c) a white telephone on a green background
   - (d) a white cross on a green background

Try this

Give examples of safety signs which are particularly appropriate to electrical installation sites.

Answer grid

1	a	b	c	d
2	a	b	c	d
3	a	b	c	d
4	a	b	c	d
5	a	b	c	d

5

Preparing and Restoring the Worksite for Electrical Installation

Remind yourself of the following important points:

Warning signs are _____ (shape), _____ (colour) and have a _____ border. What do you think they mean?

_____

_____

Mandatory signs are _____ (shape), _____ (colour) on a blue background. What do you think they mean?

_____

_____

Safe condition signs are _____(shape), _____ (colour) on a green background. What do you think they mean?

_____

Figure 5.1   Working on an industrial site.

Figure 5.2   Industrial sites can provide varying working conditions.

On completion of this chapter you should be able to:

◆ recognise different types and conditions of worksite
◆ understand the properties and strengths of different building materials
◆ identify possible disruptions that may arise from sitework
◆ recognise the need for the correct storage of components and equipment
◆ recognise the use of the different documentation required for the receipt and payment of materials
◆ understand methods of verifying circuit isolation
◆ carry out basic assessments of possible risks on site
◆ recognise the requirements for restoring the worksite
◆ complete the revision exercise at the beginning of the following chapter

Part 1

Preparing the worksite

Electrical installation work is carried out on many different types of site: domestic, agricultural, commercial and industrial. These may be in the course of construction, being extended or modified and could be empty or occupied.

The electrician's working environment will also vary greatly from a comfortable home (Figure 5.3) through to damp and noxious atmospheres, heat and cold, such as in a boiler room or chill store, to working outside in wintry conditions.

Figure 5.3   Comfort of the home?

Electrical installations involve work both inside and outside of buildings, in some cases having wiring run on the surface and in others with cables concealed under floors, in roofs and in building voids. Some wiring systems must be installed at heights, whilst others are buried in trenches.

The type of installation will depend on the construction of the building, the purpose for which it is intended and whether it is a temporary or permanent structure.

Wood, brick, steel, concrete and plasterboard are used in the construction of buildings and each of these materials has different strengths and properties which have to be recognised so that fixings and openings can be made effectively with minimum damage.

We will be finding out more about the properties of different building materials and suitable fixings in Chapter 6.

Risks

Before any work is started the site must be examined for any potential dangers.

These could include:
- insecure structures (Figure 5.4)
- inadequate lighting
- the risk of falling or being hit by falling or moving objects

- risk of drowning
- dangerous or unhealthy atmospheres
- steam, smoke or vapours.

Figure 5.4   Insecure structure?

Wherever possible risks should be removed, or access to the danger area prevented by barriers and warning notices. Consideration must also be given to the provision, wearing and use of the appropriate protective clothing and equipment.

Input services

Services such as electricity, gas, water, steam or compressed air are often present on site and the position of these must be established (Figure 5.5). The gas and water pipes will need to be earth bonded and where they prove unsuitable the deficiency should be recorded and any appropriate action taken. It must also be established that the electrical service is in a safe condition and adequate for any future additional load likely to be connected.

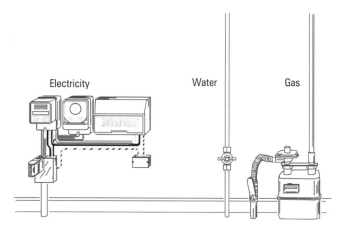

Figure 5.5   Identify input services.

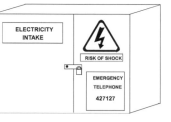

Figure 5.6   The electricity intake has been labelled and sealed.

Working at heights

Where work at heights is necessary the correct access equipment must be provided in the form of steps, trestles, ladders, scaffolding, platforms or powered mobile hoists. Warning notices and barriers are also required (Figure 5.7).

Customer disruption

Where work is to be undertaken in occupied premises it can be the cause of considerable disruption (Figure 5.8).

Figure 5.8   Least disruption?

For example, if it is necessary to disconnect power supplies to businesses or essential facilities such as hospitals, then arrangements would have to be made for work to be carried out when least disruption would be caused.

Access platforms may need to be erected overhead in a factory to install lighting equipment, or cables spread over the floor in a busy shop. In cases like these arrangements would have to be made for the work to be done either outside of normal hours or when least disruption would be caused.

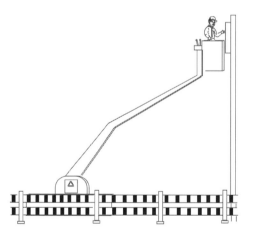

Figure 5.7   Where risks are involved warning notices and barriers should be erected.

Try this

What preparations would you consider it necessary to make if your company has been asked to provide extra power points for additional milking stalls on a local farm. The cattle will continue to be milked in existing stalls in the same area.

List potential dangers you may encounter on the following sites:

Nursery school

Brewery

Swimming pool

Give at least one possible consideration to be given before starting work in the following environments:

    domestic

    agricultural

    office

    shop

    construction site

    hospital

Dangerous risks must either be removed or

Input services such as _____, _____, water, _____ or _____ _____ are often present on site and their position established. For what purpose will gas and water pipes be required?

Part 2

Storing materials

Provision must be made so that materials and equipment such as switchgear, conduits, trunking, cable and wiring accessories can be brought onto the site as required and securely and safely stored. With larger contracts it may be necessary to have a system to ensure that the right quantities and type of materials are available at each stage of the work. Any deficiencies should be reported in good time so that action can be taken to prevent the job from becoming idle (Figure 5.9).

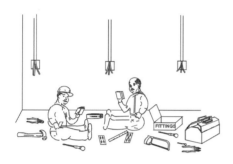

Figure 5.9   Job becomes idle!

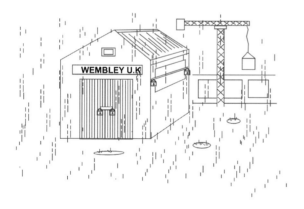

Figure 5.10   Materials need protection from theft and adverse environmental conditions.

Once the materials have been delivered and checked it is important to protect them from damage or theft (Figure 5.10). A site hut, or a room on the site which can be locked, may be required as a store for this purpose (Figure 5.11). This store must be laid out so that the materials and equipment are not damaged.

The atmosphere in the store should not be allowed to become damp in order that the condition of the equipment remains good. For example, steel conduit should be raised off the floor to prevent it rusting, or accessories kept in their wrapping to prevent discolouring due to dirt and sunlight. Some lubricants may degrade in the presence of moisture.

The protection of equipment does not end with a recognised store for even after it has been installed equipment can become damaged. It is sometimes wise to remove the inside assemblies of equipment and only fit the case while other trades are working in the area. This of course means that the inside assemblies now need even better storage facilities as they no longer have their cases for protection.

Within the storage area there may also be space for the tools and plant that is being used on the site.

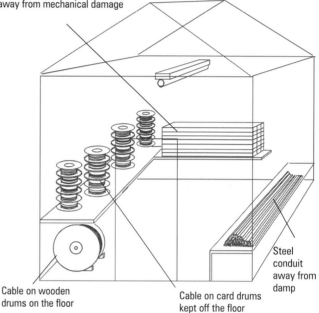

Fluorescent fittings out of any damp and away from mechanical damage

Cable on wooden drums on the floor

Cable on card drums kept off the floor

Steel conduit away from damp

Figure 5.11   The storage of electrical equipment in a site hut.

Remember

Product quality is preserved and protected by using the correct handling techniques.

Ensure that
- stock is unpacked using the correct techniques and equipment
- all packing is removed and disposed of promptly and in the correct manner
- discrepancies and/or damaged stock are set aside to be dealt with correctly
- in adverse conditions precautions are taken to prevent damage.

Instructions

Instructions as to how the work is to be carried out will depend on the size and complexity of the contract. These may vary between a simple verbal instruction and complicated written specifications, site plans and programmes involving the cooperation of customers, customers' agents and other site trades. For complex contracts a programme will be drawn up to show which trade is working on the site when, so that different trades do not get in each other's way (Figure 5.12)!

Figure 5.12   In each other's way!

Any change in the programme, for example a delay to one of the trades, will result in a knock-on effect on all the other jobs. If problems are experienced in keeping to the programme, perhaps due to sickness or bad weather, the site agent should be notified promptly so that adjustments can be made and any adverse effects minimised.

Restoring the worksite on completion

Maintaining good relations with the customer is important right through to the completion of the contract and beyond.

Figure 5.13   Restore the workplace to its original condition.

Most contracts have a date for completion. If this cannot be met for any reason then this must be taken up with the customer.

Varying degrees of "making good" are often part of the contract requirement (Figure 5.13). This can vary from simply replacing floorboards and plastering up chases to restoring original standards of decoration.

Where holes have been made for cables, conduits and trunking to pass through floors, walls and partitions which normally form part of a fire barrier, it is vital that the material removed is replaced with a fire stop of the same integrity (Figure 5.14).

Figure 5.14   Fire barriers must be replaced.

Where other structural materials must be restored it is important that the replacement materials are of an equivalent strength and characteristics to the original.

Return unused materials to stores

On completion of the work unused materials which have any value should be listed so that they can be credited back into the stores (Figure 5.15).

Figure 5.15   Return unused materials to stores.

Disposal of other waste products

Arrangements must also be made for the disposal of any waste materials. Old conduit and switchgear may be disposed of in a skip (Figure 5.16) as they do not present an environmental hazard.

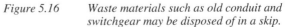

Figure 5.16   Waste materials such as old conduit and switchgear may be disposed of in a skip.

Equipment which may contain dangerous or unpleasant substances such as asbestos lagging, certain refrigerants or carcinogenic oils need specialist disposal.

Fluorescent and discharge lamps which contain highly toxic chemicals should be treated as hazardous "non-special" waste. There are lamp-crushing and cleaning systems available which minimise many of the problems associated with discharge lamp disposal. Figure 5.17 shows the cross-section of such a machine.

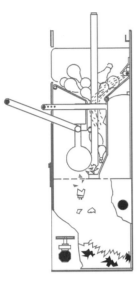

Figure 5.17   *Lamp crusher.*
Reproduced with kind permission of
BALCAN ENGINEERING LIMITED, LINCOLNSHIRE.

Figure 5.18   Reproduced with kind permission of
BALCAN ENGINEERING LIMITED, LINCOLNSHIRE.

Points to remember ◄ – – – – – – – – – – – – – –

Storing materials

A system should be instituted to protect materials and components from damage, during transit and storage or from deterioration, loss or incorrect identification.

1. On arrival items should be verified for damage and compliance with requirements.
2. Components should be identified with job/customer and possibly drawing number.
3. The storage area should be as clean and dry as possible, secure and with access only to authorised staff.
4. There should be a procedure for the issue of stores to prevent loss, incorrect use and the mixing of stock.
5. There should be frequent checks to detect any loss or degradation of stock.

Restoring the worksite

At the completion of the work the site should be restored to an equivalent condition, this could mean
- replacing floorboards
- redecorating walls, skirting boards etc.
- where fire barriers have been breached they must be replaced with a fire stop of the same integrity
- removing unused materials and crediting them back into the stores
- disposing waste materials in the most appropriate way:

    fluorescent lamps should be

    old switchgear may be

    old conduit may be

Self-assessment multi-choice questions
Circle the correct answers in the grid below.

1. Powered mobile platforms may only be operated by persons over
   - (a) any age
   - (b) 16 years
   - (c) 18 years
   - (d) 21 years

2. Cable on card drums should be stored
   - (a) on the ground
   - (b) outside
   - (c) raised off the floor
   - (d) in the site office

3. Materials received on site should be signed for on
   - (a) a delivery note
   - (b) a time sheet
   - (c) an invoice
   - (d) a requisition form

4. Fire barriers must be replaced when a trunking run
   - (a) changes direction
   - (b) is terminated
   - (c) enters a distribution board
   - (d) passes through a structural brick wall

5. Fluorescent lamps are best disposed of in
   - (a) a skip
   - (b) a lamp crusher
   - (c) the local bottle bank
   - (d) a dustbin

Answer grid

1	a	b	c	d
2	a	b	c	d
3	a	b	c	d
4	a	b	c	d
5	a	b	c	d

Progress check

1. What is a permit-to-work?
   (a) A permit which lets you work on any equipment you have been trained to use.
   (b) A permit which says you can work on the equipment as long as it is dead, and as long as you have the key.
   (c) A permit which tells you what work to do, what precautions you should take and which provides a key for locking off.
   (d) A permit which says what machine to work on, the tools you should use and gives you a key to switch it off.

2. If the fire alarm has been initiated at your workplace what is the first thing *you* should do?
   (a) telephone the fire brigade
   (b) check to see if the alarm is genuine
   (c) wait for your supervisor to tell you what to do
   (d) evacuate the building immediately

3. What does "RCD" stand for?
   (a) residual current deactivator
   (b) residual current device
   (c) residual circuit deactivator
   (d) residual circuit device

4. What colour would a 415 volt connector be under BSEN60309-2:1992?
   (a) red
   (b) blue
   (c) white
   (d) light green

5. What fire extinguisher can be only be used on Class A fires?
   (a) CO_2 gas
   (b) water
   (c) foam
   (d) dry powder

6. A person turns up at your work site stating that he is a health and safety inspector. What should you do?
   (a) escort him around the site so that he can see what he wants
   (b) go and inform your supervisor
   (c) verify the person's identity before proceeding
   (d) go and check to see that the person has made an appointment

7. Before using a portable electric tool what should you do?
   (a) switch off before you plug in
   (b) check that it has a label stating that it has been tested
   (c) visually check it for damage
   (d) make sure the voltage of the equipment matches the supply

8. If you see a colleague not wearing ear protection in an ear protection zone what should you do?
   (a) tell your colleague that you are going to inform the supervisor
   (b) tell your colleague to wear his ear protection
   (c) do not do anything as it is not your responsibility
   (d) tell your supervisor straight away

9. Under what situations is someone allowed to work on live electrical equipment?
   (a) if they have been told to do the work and have warned other people to stay away
   (b) if they have been told to do the work and are fully trained
   (c) if the work can be done no other way and they are competent enough to carry out the work
   (d) no one is allowed to work under live conditions by law

10. An employer is required by law to provide
    (a) hand washing facilities and free laundry
    (b) tea facilities and a reasonable working temperature
    (c) a safe work method and training in its use
    (d) accident reporting procedure and weekly safety reports

11. Who has the legal powers to prosecute an employer for not complying with the Health and Safety at Work Etc. Act 1974?
    (a) an employee
    (b) a safety representative
    (c) shop foreman
    (d) The Health and Safety Inspectorate

12. On a construction site or in a factory the voltage used on hand held portable equipment is restricted above earth potential to
    (a) 55 V a.c.
    (b) 110 V a.c.
    (c) 120 V a.c.
    (d) 240 V a.c.

13. Which of these should *not* be used to pull a casualty clear of a live electrical conductor
    (a) rubber gloves
    (b) rope
    (c) wooden pole
    (d) metal ladder

14. On which of the following types of fire would you use a fire extinguisher with a red panel (water)?
    (a) wood
    (b) hot fat
    (c) electrical
    (d) gas

15. A written health and safety policy must be provided by an employer whose employees number
    (a) 15 and over
    (b) 10 and over
    (c) 5 and over
    (d) 2 and over

16. When materials that have been ordered are received on site they will be accompanied by
    (a) an order
    (b) a delivery note
    (c) a requisition
    (d) a job sheet

17. When using power tools on site the recommended voltage would normally be
    (a) 400 V
    (b) 230 V
    (c) 220 V
    (d) 110 V

18. The time of day when the demand for electricity is usually at its highest is
    (a) 3 a.m.
    (b) 6 a.m.
    (c) midnight
    (d) 6 p.m.

19. The maximum frequency the supply company can legally supply when it is declared as 50 Hz is
    (a) 56 Hz
    (b) 52.5 Hz
    (c) 51 Hz
    (d) 50.5 Hz

20. The isolators in a high voltage switching station are for
    (a) breaking the circuit in overcurrent conditions
    (b) switching off the load to the generator
    (c) switching off the load to the grid lines
    (d) opening the circuit after the circuit breaker has switched off the load

Answer grid

1	a b c d		6	a b c d
2	a b c d		7	a b c d
3	a b c d		8	a b c d
4	a b c d		9	a b c d
5	a b c d		10	a b c d
11	a b c d		16	a b c d
12	a b c d		17	a b c d
13	a b c d		18	a b c d
14	a b c d		19	a b c d
15	a b c d		20	a b c d

6

Building Materials and Fixing Components

Remind yourself of the following important points:

Fluorescent fittings should be stored _____.

Steel conduit should be stored _____.

Cable on _____ drums can be stored on the floor.

Accessories should be kept in their wrapping to prevent _____.

Where holes have been made through floors or walls for cables, conduit or trunking the material removed must be replaced by material of _____.

If the wall is a fire barrier, material removed must be replaced by a fire stop.

On completion of this chapter you should be able to:

◆ understand basic types of building structure and the materials used in them
◆ identify the following fixing methods:

    screwing
    bolting
    anchoring
    plugging

◆ recognise different tools for making fixings and determine their suitability
◆ complete the revision exercise at the beginning of the following chapter

Electrical installation work has to be carried out in buildings which will vary greatly in their age, construction and use.

Fixings for conduits, cables, accessories and equipment need to be made both internally and externally and to a variety of materials.

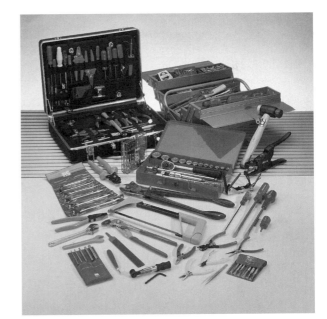

Figure 6.1   Tool kits.

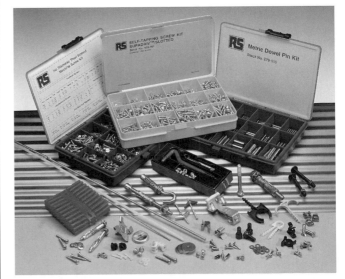

Figure 6.2   Fasteners and fixings.

Part 1

Building materials

Stone

Natural stones such as granite, limestone, sandstone and slate are used in building construction, the hardest and most durable being granite. Artificial stone, manufactured from cement and natural aggregate is used as a "cheaper" alternative to natural stone.

Bricks

Bricks are made from clay, concrete or calcium silicate under three broad classifications:

- common
- facing and
- engineering

All of these can vary in their "hardness" and durability.

Bricks can be (Figure 6.3):

- solid
- perforated or
- "frog"

and the type of brick used can affect the selection of the method used to make fixings.

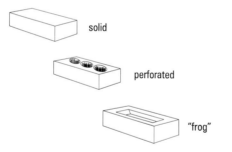

Figure 6.3   Types of brick.

Blocks

Building blocks cast in concrete are available in a range of sizes with various densities.

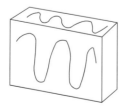

Figure 6.4   Aerated block.

Aerated blocks (Figure 6.4) are widely used for internal partitions. They are lightweight and can be easily chased to take conduit.

Dense (Figure 6.5) and lightweight solid blocks are also widely used and some types can be load bearing.

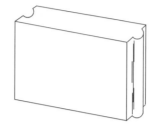

Figure 6.5   Jointed dense block.

Hollow blocks (Figure 6.6) have cavities which can be filled with concrete or fitted with mild steel strengthening rods.

Figure 6.6   Hollow block.

Cellular blocks have cavities which don't go right through the block.

Reinforced concrete

Instead of using brick or block, some buildings have a frame of reinforced concrete or steel on which is hung a cladding material such as aluminium, wood, plastic or glass. The concrete is usually made by mixing portland cement with aggregates and in load bearing structures may use steel reinforcement.

Ceramic materials

In kitchens and bathrooms many surfaces are covered with glazed wall tiles. These are extremely hard and durable but so brittle that care has to be taken when drilling the tile.

Wood

Many fixings are made on wood as it is still used extensively in buildings with hardwoods or softwoods used in construction. Chipboard, which is manufactured from chips of wood or shavings mixed with synthetic resin glue, is used for flooring or panelling when it is well supported. Hardboards and fibre building board are often used in internal walls, partition walls or for lining ceilings.

Stud partition walls

Timber framed partition or "studding" walls (Figure 6.7) are often used to make internal partitions. These consist of a "head or top plate" fixed to the ceiling, a "sole plate" attached to the floor with vertical "studs" between them.

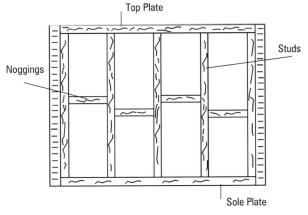

Figure 6.7   Stud partition wall.

Sheets of plasterboard are simply nailed to the framework to complete the wall.

Wood framed buildings

Some very old buildings were constructed with a wooden frame with its spaces filled with mortar and bricks.

Modern buildings utilise a timber frame construction (Figure 6.8) with plywood, a plasterboard inner wall with an outer skin of bricks.

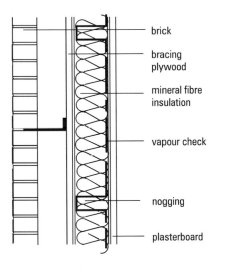

Figure 6.8   Modern timber frame construction.

New constructions will usually have working drawings which show exactly what type of partition wall you will find when you install an electrical installation. It is important to know about the construction and the materials used in order to make secure fixings.

Try this

Ask around your friends and colleagues and list all the different structures of the buildings in which they live or work. Note any particular problems that may be encountered when fixing conduits, cables or accessories to these structures.

Buildings are made from many different materials and it is important to understand what they are constructed from before commencing an installation. Some of the most common materials are:

Concrete is usually made by mixing _____ and _____.

In load-bearing structures what form of reinforcement may be used?

When drilling kitchen or bathroom tiles why is special care required?

Timber framed partition walls consist of (name the parts indicated in Figure 6.9):

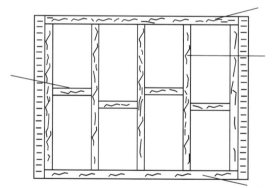

Figure 6.9

What is nailed to the framework to complete the wall?

How might you find out what type of partition wall has been used in a new construction?

Part 2

Fixings for wood, masonry and metal

Nails

Nails are commonly used for fixing wood to wood. Grip is provided by friction derived from the compressive reaction of the fibres in the wood.

Masonry nails

Masonry nails are intended to be driven directly into brickwork without any need to drill or plug. They are very hard and, as with many fixing operations, eye protection from flying chips of metal or masonry is required. They are not considered suitable fixings for accessories.

Woodscrews

Although woodscrews (Figure 6.10) take longer to put in they are more effective than nails and have the advantage that they can be removed and put back again. Plastic masonry plugs are used to help secure screw fixings.

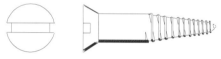

Figure 6.10   *Countersunk screw.*

Twin thread woodscrews (Figure 6.11) have the advantage that they can be driven in twice as quickly as common woodscrews because they have two separate parallel helixes.

Figure 6.11   *Twin thread screw.*

Standard screws are made from mild steel, which can rust, so for certain locations zinc-plated or brass screws are used.

For heavier loads a coach screw (Figure 6.12) can be used, which has a square head so that it can be tightened using a spanner.

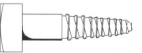

Figure 6.12   *Coach screw.*

Hammer screws

A development of the plastic masonry plug is the hammer screw (Figure 6.13). Here the hole is predrilled and the plug inserted, then the screws are tapped home with a hammer. The nylon plug allows the screw to slip in, then springs back to grip the threads. The screw can be taken out in the normal way.

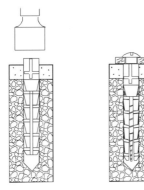

Figure 6.13   Hammer screw.

Expanding anchors

There are many different types of expanding anchor (Figure 6.14), which are derived from two basic principles. In the first a bolt is fitted inside an expanding sleeve. As the nut is tightened a cone separates the metal sleeve so that the sides of the hole are gripped. In the second type the anchor is "set" by power hammering the sleeve so that it expands inside the hole. Most types of metal anchor are made in either galvanised or stainless steel.

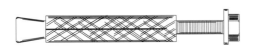

Figure 6.14   Expanding anchor.

Chemical anchors

As an alternative to plugging, particularly where the fixing is to be made into difficult materials, hollow masonry, or near outside edges, chemical or resin anchors (Figure 6.15) can be used. A mixture of resin and hardener is injected into the hole and a bolt or special threaded sleeve inserted. The fixing can carry a load as soon as the mix has "cured".

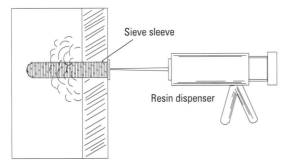

Figure 6.15   A chemical anchor using a sleeve in hollow masonry.

Thin wall fixing

Fixings to ceiling and walling materials such as plasterboard, hardboard and chipboard cannot normally be made by nailing or screwing. Instead, toggle fastenings or "cavity wall anchors" (Figure 6.16) must be used. These have a flange at the front and a split sleeve at the back which collapses as the screw is tightened.

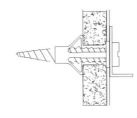

Figure 6.16   Cavity wall anchor.

Other types of plasterboard anchor (Figure 6.17) are also available. The fixings feature a deep thread and are made of plastic or a zinc alloy. The plug is screwed in first, sometimes requiring a setting tool, and then the fixing is completed with a screw.

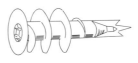

Figure 6.17   Plasterboard anchor.

For switches and switchplugs on plasterboard or cavity partition walls it is common practice to use dry lining boxes. These have twin spring-loaded fixing lugs which clamp the box into the aperture.

Clips and clamps

For structural reasons drilling is often out of the question on steel girders so clamped fixings (Figure 6.18) can be used, which save both time and labour costs. They are easy to fix and very often the only tools required are a hammer, screwdriver and spanner. They simply clip on to the building structure, and the item being installed can be snapped into place.

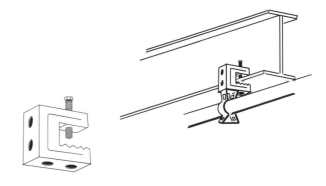

Figure 6.18   Clamp fittings.

Shot fired fixings

Shot fired fixings are nails or threaded studs that are driven into concrete, brick or even steel joists using a tool with explosive cartridges. The tool cannot be fired unless the spring loaded barrel is firmly pressed against the work surface. When properly used by a trained, qualified and competent person this method of making fixings into very hard materials is both rapid and safe.

Cartridge tools must only be used by trained and competent operators.

DON'T use cartridge tools unless you are properly accredited.

Specifications often state that cartridge tools must not be used.

Tools for making fixings

There is a wide range of tools both manual (Figure 6.19) and powered which can be used for making fixings into different materials.

Figure 6.19   Hand drive tool for setting nails into concrete.

Powered tools are now commonly used, and for complete portability on sites where no electricity is available battery-powered drills are increasingly common. These are fitted with adjustable drilling depth stops to save battery power.

Rotary hammers (Figure 6.20) are used to make light work of drilling and anchor setting and these may be cam or electro-hydraulic action.

Supplies on site should preferably be 110 V derived either from an isolating transformer or a portable generator.

Suitable ear and eye protection must always be worn when fixings are being made.

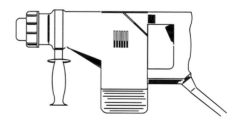

Figure 6.20   Rotary hammer.

Remember

Employers must (BY LAW)

- provide and maintain safe tools for use by their employees
- provide any necessary training in their use
- provide information regarding safety procedures
- provide supervision where necessary
- ensure the work method is safe.

Employees must (BY LAW)

- take reasonable care for their own safety
- not endanger others
- cooperate with their employer on safety procedures
- not interfere with tools etc. provided for their use
- correctly use all work items provided in accordance with instructions and training given to them.

Try this

You have been sent to a job without the most suitable fixing materials. Obviously someone in your company has not assessed this, or has failed to correctly communicate the situation before you were sent out. What should you do about this?

Table 6.1 shows some of the more common materials and some fixings suitable for the type of material used.

Table 6.1

Material Type	Fixings	Suitable Tools
Brickwork Common or facing bricks are made from clay. They are fairly easy to drill and have good fixing properties. Engineering bricks are harder and fairly difficult to drill.	Screw fixings Fibre or plastic plugs Masonry nail	Hand drill Electric drill Rotary hammer drill Hammer
Concrete Made from cement and aggregate in dense and lightweight forms as for concrete blocks **Concrete Blocks** Lightweight (clinker) concrete blocks are easy to drill and have fair fixing properties. Dense (limestone) concrete blocks are fairly easy to drill and have good fixing properties.	Screw fixings Fibre or plastic plugs Bolts – several different methods Resin anchors Stud anchors	Hand Drill Electric drill Rotary hammer drill
Steel	Snap on fixings, girder clips Shot fired cartridges	Hammer, screwdriver and spanner Cartridge powered fixing tool
Plasterboard Lath and plaster Hardboard, plywood, fibre board and chipboard Difficult to get good fixings	Spring toggle fasteners Gravity toggle Cavity sleeve Cavity wall anchor Dry lining boxes	Drill Screwdriver
Wood Easy to drill Good fixing properties	Wood screws Nails	Drill Screwdriver Hammer

A basic tool kit required by an electrician

A basic tool kit (Figure 6.7), which should built up by all trainee electricians, would include such items as

for fixing cable
- pin hammer
- claw hammer
- screwdrivers for slotted head and cross head screws
- junior hacksaw
- spanners
- spirit level
- plumb-bob and line
- metric rule
- bolster cold chisel
- club hammer

for stripping and terminating the cable
- knife (not Stanley type)
- side cutters
- strippers
- electrician's terminal screwdrivers for lighting and power connections
- pliers
- small file
- spanners

As your career progresses you will obviously acquire more tools to meet your needs. You should purchase a secure toolbox to put your own hand tools in. Remember that damaged tools must be repaired or replaced for your own and others' safety.

Tools for testing the cable and specialist items, such as an adhesion gun, may be provided by your employer.

Tools are expensive to replace. Take care of your own and your employer's tools, and keep them clean and in good condition. It may also be a good idea to consider insuring your equipment.

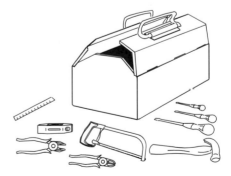

Figure 6.21   Basic tool kit.

Points to remember ◀ – – – – – – – – – – – –

Before carrying out any fixing work it is important to note what material the building has been constructed from. This will determine the type of fixings and equipment to be used.

Building materials have different strengths and characteristics. The most common are _____, _____, plaster, _____, steelwork and concrete.

Generally there is a suitable fixing method available for every type of construction fabric found on site.

The different fixing methods fall into six main categories:

> plugging
> bolting
> screwing
> clamping
> anchoring
> adhesives

The equipment which can be used for achieving fixing can range from hammers and screwdrivers through to adhesion guns and cartridge hammers.

Try this

Using the list of types of surface to which fixings for electrical components can be made (on page 67), obtain a manufacturer's catalogue to help you suggest the most suitable fixing method for each surface, note the part/order code number and suggest suitable tools for the work.

Self-assessment multi-choice questions
Circle the correct answers in the grid below.

1. The fixing device shown in Figure 6.22 is

Figure 6.22

   (a) a coach screw
   (b) a countersunk screw
   (c) an expanding anchor
   (d) a hammer screw

2. A toggle fastening is used for a fixing in
   (a) plasterboard
   (b) brick walls
   (c) steel girders
   (d) masonry

3. On a timber framed partition the horizontal piece of timber which forms the base of the structure is called a
   (a) top plate
   (b) nogging
   (c) stud
   (d) sole plate

4. Chemical anchors can be used where
   (a) the material is soft
   (b) the environment is corrosive
   (c) the material is brittle
   (d) the fixings are to be made near edges

5. The fixing shown in Figure 6.23 is

Figure 6.23

   (a) a coach screw
   (b) a countersunk screw
   (c) an expanding anchor
   (d) a hammer screw

Answer grid

1	a	b	c	d
2	a	b	c	d
3	a	b	c	d
4	a	b	c	d
5	a	b	c	d

7

Measuring and Marking Out

In the last chapter we looked at the materials used in buildings and the fixings that can be used. Check that you can remember the following points:

Building blocks are generally cast using _____.

Three common types of brick are

Concrete is usually made of _____ and _____.

_____ are commonly used to fix wood to wood.

How does a hammer screw work?

Figure 7.1   Rules and measuring tapes.

What fixings can be used on steel girders to save both time and labour costs?

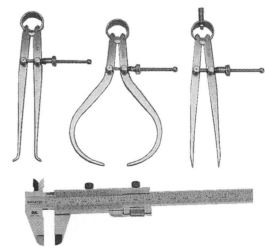

- ◆ take measurements using scale rules
- ◆ recognise the instruments needed to take measurements on round objects
- ◆ identify marking out equipment for defined situations
- ◆ transfer information from drawings
- ◆ identify block, circuit and wiring diagrams
- ◆ complete the revision exercise at the beginning of the following chapter.

Figure 7.2   Callipers and Vernier scale.

Part 1

Measuring

The penalty spot on a soccer pitch (Figure 7.3) should be 12 yards (11 m) from the goal line. This could be determined by pacing it out, and although this measurement could be quite accurate, it may actually be some distance out. The degree of accuracy required in marking the penalty spot may depend on the importance of the match to be played. Similarly, all measurements are related to the accuracy required for their purpose.

Measurement can be carried out using many different devices and the selection of the correct one may depend on factors such as the accuracy required, the shape of the object and positioning, among many others.

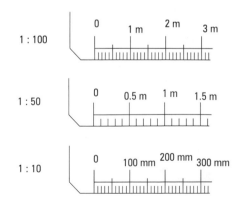

Figure 7.3

Rules

By far the most used device for measuring is the rule. However, there are probably more different types of rule in use than any other measuring instrument. The length of a rule can vary from a few centimetres to many metres long, and the graduations will also vary depending on what the rule was designed for.

A standard office rule is about 300 mm long, made of moulded plastic with one side divided into mm and cm sections. A similar looking rule may be used in a drawing office, but the graduations would show scales rather than millimetres and centimetres. Figures 7.4 and 7.5 compare some of the common scales in use with a standard centimetre/millimetre rule.

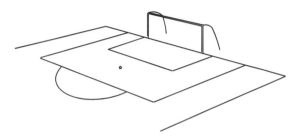

Figure 7.4   Scale rules.

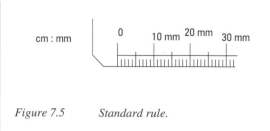

Figure 7.5   Standard rule.

Tapes

Where longer distances are involved measuring tapes are used. A tape consists of a flexible measuring strip which is kept rolled up in a case. To take measurements the strip is pulled out; it is then rolled back into the case after use. Tapes up to 6 m are usually spring loaded (Figure 7.6) so that they automatically return into their case after use. Longer tapes are fitted with a fold-away handle to manually rewind the measuring blade (Figure 7.7).

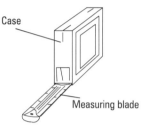

Figure 7.6   Pocket tape for measuring up to about 3 m.

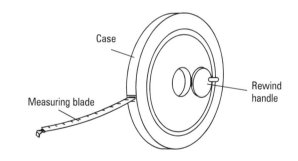

Figure 7.7   Long tape with handle to rewind tape after use.

For a quick assessment of longer distances, for example the dimensions of a room, an ultrasonic distance measuring device can be used.

Callipers

Some shapes are very difficult to measure with a rule or tape. In these cases other devices have to be used to convert measurements into a linear shape. For example, it is sometimes difficult to determine the outside diameter of a fixed conduit just by looking at it, especially if it may be confused between imperial and metric dimensions.

Using "outside" callipers the diameter of the conduit can be set as shown in Figure 7.8. The callipers can now be laid against a rule and the outside dimension of the conduit can be determined as in Figure 7.9.

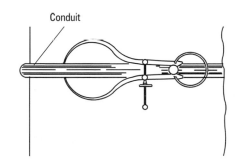

Figure 7.8   Outside callipers.

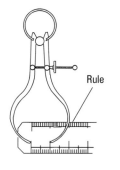

Figure 7.9

Similar measurements can be taken inside a pipe using "inside" callipers (Figure 7.10).

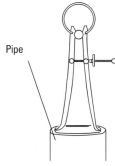

Figure 7.10   Inside callipers.

An alternative to using callipers would be to use a Vernier scale (Figure 7.2).

Cable gauges

Where the diameter of conductors has to be measured none of the methods discussed so far is accurate enough. Conductor sizes are usually listed in tables by their cross-sectional area. As most of the conductors we use are round there is a direct relationship between the diameter and the cross-sectional area. To save a great deal of calculation, cable manufacturers have gauges (Figure 7.11) for the common sizes of conductor.

The gauges are really measuring the diameter of the conductor, but the size shown is given in cross-sectional area.

The conductor is placed into the slots and when it just goes in without slopping about, that is the size of the conductor.

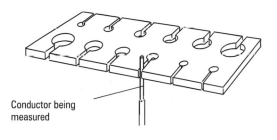

Figure 7.11   Cable gauges.

Transferring information to site

Electrical designers usually detail the position of electrical accessories and equipment on scaled drawings. This means that a comparatively small sheet of paper can represent a large area with a lot of electrical equipment in it. To transfer the information from the drawing to the actual area the scale of the drawing must be known. Then, using a scaled rule on the drawing, the actual measurement can be determined (Figure 7.12). A tape can then be used on site to measure the exact position as shown in Figure 7.13.

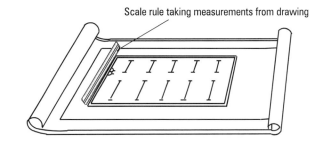

Figure 7.12   A scaled drawing.

Figure 7.13 shows how measurements can be transferred to wall positions, but it is not always so easy when lights have to be marked out, particularly as a one-person operation.

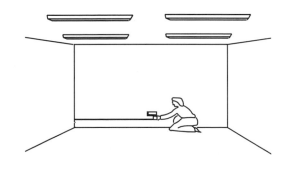

Figure 7.13   Measurement being transferred on site

One of the best ways to overcome the problems of holding tapes at ceiling level is to mark out the lighting points on the floor. Using a plumb line the point on the floor can be transferred to the ceiling (Figure 7.14).

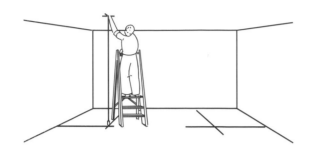

Figure 7.14

Chalked lines

Where long straight lines are required, the plumb line, or another line, can be used. The line should be covered in chalk and then with each end of the line anchored it can be "pinged" (Figure 7.15) so that the chalk is transferred to the surface. This method can be used for vertical or long horizontal runs.

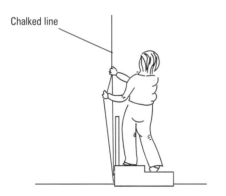

Chalked line

Figure 7.15

In some circumstances it is not possible to use this "chalk line" method of marking out. In such cases a spirit level and straight edge have to be used. The spirit level (Figure 7.16) is a very helpful aid when installing any vertical or horizontal equipment, but it must be looked after. If the edges of it become damaged or the spirit tube becomes loose, the accuracy of the level will be affected.

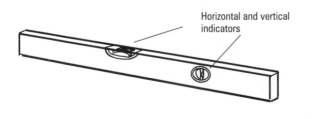

Horizontal and vertical indicators

Figure 7.16   Spirit level.

Water levels (Figure 7.17) are used to transfer a horizontal level from one location to another. This device is based on the principle of water finding its own level.

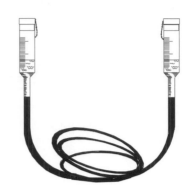

Figure 7.17   Water level.

Points to remember ◀ – – – – – – – – – – – – – – – –

The device most used for measuring is the rule. In a drawing office it would be a _____ rule.

Where longer distances have to measured a _____ may be used.

The outside diameter of a conduit would be measured using _____ and _____.

Cable gauges measure the diameter of the conductor but the size given is the _____.

Describe a way of transferring measurements for lights on the ceiling from a scaled drawing.

If it is not possible to use a "chalk line", how can vertical or horizontal lines be drawn?

Try this

Accurately measure the room you are working in and draw it below to a scale of 1:50. Mark in the position of any doorways or windows. Make a list below the diagram of all the electrical lighting equipment in the room – you will need this information later.

Part 2

Marking out

When marking out on different surfaces the correct tools should be used. If a cross has to be marked on a brick wall then chalk or crayon can be ideal. However, if an accurate mark has to be made on a metal surface then a scriber (Figure 7.18) and centre punch may be required.

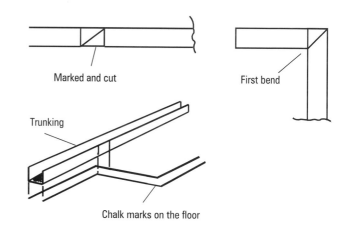

Figure 7.18   A scriber.

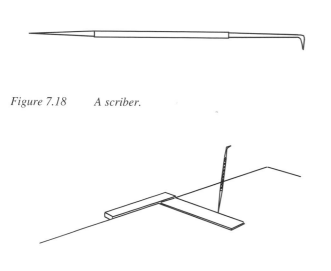

Figure 7.19   A scriber used with a square on a metal surface.

A square, as in Figure 7.19, is used for measuring and checking angles of 90°. It comprises two parts, the stock and the blade, and it is very accurate.

A scriber can also be used with a straight edge, a heavy duty steel or alloy strip along which the scriber can be drawn to mark metal or other surfaces.

Metal surfaces that are to be drilled should first have the position marked out with a scriber. Then a centre punch (Figure 7.20) should be used to create an indentation that the drill will locate in. This will reduce the possibility of the drill sliding across the smooth metal surface and ensure that the hole is drilled in the correct position.

Figure 7.20   A centre punch.

The electrician is often going to be involved with working on metal surfaces. These surfaces may be parts of distribution boards and bus-bar chambers, or on installation equipment such as trunking and cable tray.

Taking a section of trunking as an example, changes in direction will often have to be fabricated on site. This means that in addition to the normal site tools, equipment such as squares, scribers and centre punches is also required. The fabrication of a trunking bend to fit a site requirement can be quite complex and if care is not taken costly material can be wasted. It is sometimes a good idea to chalk out the shape of the finished bend on the floor (Figure 7.21) or some other surface which will not be damaged. If this is carried out carefully, and to scale, the measurements can be transferred directly to the trunking by laying the trunking onto the drawing.

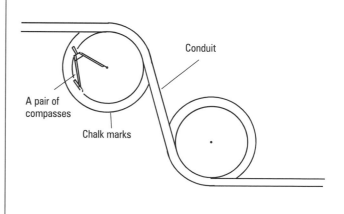

Marked and cut

First bend

Trunking

Chalk marks on the floor

Figure 7.21

A similar process may be used when conduit shapes are to be made, but compasses may be necessary to get the exact shape required (Figure 7.22).

Conduit

A pair of compasses

Chalk marks

Figure 7.22

Standard symbols

Whenever the location of electrical equipment has to be shown on plans and drawings there are standard symbols that should be used. Some of these have been used already in this section.

single-pole, one-way switch

lighting point or lamp
(general symbol)

main control or intake point

Figure 7.23

Below are listed a few more that are likely to be encountered, all of which are taken from British Standard 3939.

cord operated single-pole,
one way switch

two way switch

intermediate switch

socket outlet
(general symbol)

switched socket outlet

fluorescent lamp - single tube

distribution board
(general symbol)

Figure 7.24

Drawings and diagrams

The most frequently used diagrams that we will come across include:

- block diagrams
- circuit diagrams
- wiring diagrams
- layout drawings

Block diagrams

These indicate the sequence of components or equipment. Each item is represented by a labelled block (Figure 7.25).

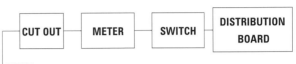

Figure 7.25   Block diagram.

Try this

If you are using a scale of 1:50 and the lines on the drawings measured the following what would be the represented length?

1 cm

5 cm

25 cm

3 cm

6.5 cm

If the following measurements are taken on site and are being put on a scaled drawing of 1:10 what would the length of the measurements on the drawing be?

300 mm

1 m

20 cm

2.8 m

Try this

On the requisition form make a list of materials required for the installation in Figure 7.26. Assume all cables and fittings are to be mounted on the surface.

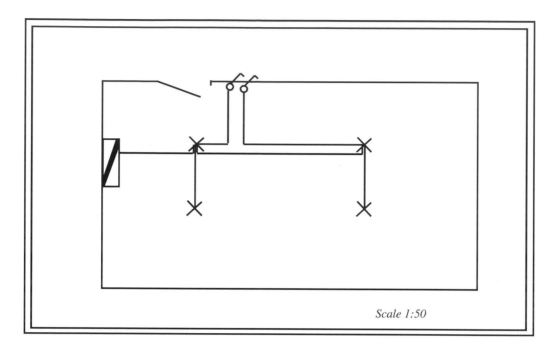

Scale 1:50

Figure 7.26

Symbols one-way switch – 1.3 m from ceiling height

✕   ceiling light – each consisting of: a 3-plate ceiling rose
0.3 m of twin heat resisting flex
a heat resisting lamp holder

consumer's unit – 4-way MCB consumer unit 1 m from ceiling height

On the Requisition Form below make up a list of materials required for this installation.
Assume all cables and fittings are to be mounted on the surface.

REQUISITION FORM	JOB TITLE		NO.
Description	Quantity	Catalogue No.	Notes

Circuit diagrams

Circuit diagrams (Figure 7.27) are used to show how the components of a circuit are connected together. The symbols represent pieces of equipment or apparatus and the diagram will show how the circuit works.

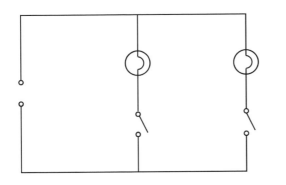

Figure 7.27   Circuit diagram.

Wiring diagrams

Wiring diagrams (Figure 7.28) indicate the locations of the components in relation to one another and cable connections, and are more detailed than circuit diagrams.

Layout drawings

Layout drawings (Figure 7.29) are used to indicate the particular location of outlets, accessories, components etc. The drawings themselves will be to scale and it is usual for the design engineer responsible for the installation to issue them.

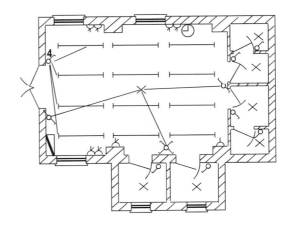

Figure 7.29   Layout diagram.

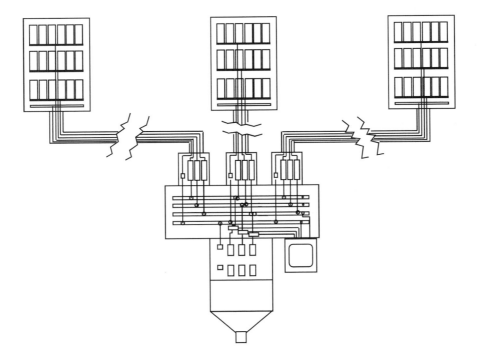

Figure 7.28   Wiring diagram.

As fitted drawing

Another type of drawing is the "as fitted" drawing (Figure 7.30). This shows the exact location of cable runs, the method of installation and the final position of all items of equipment. They are drawn on site either on overlays or copies of the original site layout drawings. They are usually kept up-to-date by the site foreman, and when the work is complete a copy will be handed over to the client's representative.

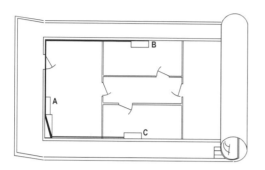

Figure 7.30   *As fitted drawing showing sub main cables going from the main intake position to the distribution boards A, B and C.*

Points to remember ◄ – – – – – – – – – – – – – –

The accuracy of measurements depends on the application they are required for. Transferring measurements accurately from scaled drawings to site is important if the original design is going to be implemented.

Measurements can be made using other equipment apart from rules. Some, such as callipers, are used in conjunction with rules. The method used for marking out depends on the surface being used and the accuracy required. If holes are to be drilled then the holes should be marked with a centre punch.

The interpretation of all types of diagrams is a vital part of the electrician's job.

Block diagrams are labelled _____ showing the sequence in which items are connected.
_____ diagrams show the electrical operation of a particular circuit or system.
Wiring diagrams are used to show the actual connections involved and the relative component locations.
Layout drawings show the position of components and equipment and will be drawn to _____.

Circuit and layout drawings should use the appropriate BS_____ symbols.

Try this

On the plan drawn on page 73 use the relevant symbols to show the position of electrical lighting equipment.

Self-assessment multi-choice questions
Circle the correct answers in the grid below.

1. A scale of 1:50 is used on a drawing. This means that 1 cm represents
   (a) 0.5 m
   (b) 1.0 m
   (c) 2.0 m
   (d) 50 m
2. A scriber can be used to
   (a) make an indent for drilling a hole
   (b) draw curves on wooden surfaces
   (c) work out scales on a drawing
   (d) mark lines on metal surfaces
3. A cable gauge is labelled "cross-sectional area of conductor". The measurement gauged is actually the
   (a) cross-sectional area
   (b) radius
   (c) diameter
   (d) circumference
4. The symbol shown in Figure 7.31 is that for a

Figure 7.31
   (a) one-way switch
   (b) socket outlet
   (c) lighting point
   (d) distribution board
5. Callipers can be used to
   (a) mark out lines on metal surfaces
   (b) measure cable runs
   (c) measure the diameter of conduits
   (d) gauge the cross-sectional area of conductors

Answer grid

1	a	b	c	d
2	a	b	c	d
3	a	b	c	d
4	a	b	c	d
5	a	b	c	d

8

Moving Loads

Check that you can remember the following.

Label the circuit symbols shown below:

 _____

 _____

_____

_____

_____

Figure 8.1 Sack trolleys for manual handling.

Figure 8.2 An electrically powered hoist.

Figure 8.3 Hydraulic jacks.

On completion of this chapter you should be able to:

◆ identify the method of moving a load related to its mass
◆ recognise the correct way to manually lift a load
◆ recognise the basic principal of the lever
◆ identify where pulleys should be used
◆ identify the safety requirements when using slings and pulleys
◆ recognise where barrows and trolleys can be used
◆ complete the revision exercise at the beginning of the following chapter

Figure 8.4 A diesel generator could provide the power for a hoist.

Part 1

Manual lifting

What is a load?

A load is an object which has to be moved or lifted. One example of this could be where 100 m reels of 1.5 mm^2 cable have to be moved from the floor on to a work-bench (Figure 8.5).

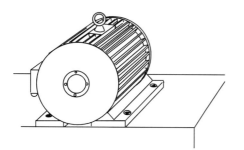

Figure 8.5

Another example is a heavy electric motor which has to be moved from the stores into the workshop area (Figure 8.6).

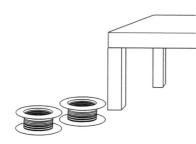

Figure 8.6

Both the above examples require a load to be moved from one place to another, but the methods used to achieve this would be very different.

Manually handling loads

It is important to recognise that moving a load can involve a number of different considerations. The Manual Handling Operations Regulations 1992, which came into force on 1 January 1993, recognise the possible risks involved in moving loads. By working through the flow charts and check lists shown in the Guidance on the Regulations it is possible to limit the risk to injury when carrying out manual handling operations. Many manual handling injuries arise from repeated handling using the wrong technique or incorrect posture. This can even lead to permanent disability.

When considering moving a load the question should be asked:
- Is it necessary to move the load?

If yes:
- assess the task and then
- reduce the risk of injury to the lowest reasonably practicable level, which may require the use of mechanical assistance.

It is generally accepted that loads over 20 kg need lifting gear, so remember – if it is appropriate to use equipment, it has been provided and you have been trained to use it then it is your *legal* requirement to do so.

Guidelines

In general there are a number of factors that must be considered when a load has to be moved which include:

The task

Does it involve moving the load in a way that may create a risk, for example does it mean that reaching up or a twisting movement is involved (Figure 8.7)? Are excessive distances involved, is there a risk of sudden movement of the load and will there be sufficient rest periods?

Figure 8.7

The actual load

Could the load itself cause injury?

How can you find out how heavy it is? Is there any documentation with it or a label on it or can you estimate how heavy it is by looking at it?

Is it a bulky or difficult shape to get to grips with (Figure 8.8)? Some objects may be greasy or wrapped in loose packing and these may need extra care. Has it got suitable handling points where you can get hold of the object? Has it any sharp corners or is it hot?

Figure 8.8

The working environment

Is the route free from obstacles (Figure 8.9) or what preparation needs to be carried out to clear it?

Is the destination clear and does anything have to be arranged before starting off with the load?

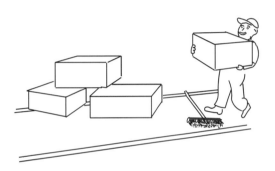

Figure 8.9

Will there be enough headroom or will you have to stoop, and are there any fixtures (or furniture) in the way?

Is there any likelihood of an uneven or slippery floor?

Will you be carrying a bulky light load and be caught by a sudden gust of wind?

Consideration must also be given to factors such as the structure of the building, for example uneven floor levels, the temperature, humidity and level of illumination.

The capability of the staff involved

Are the staff trained to carry out the task?

Consideration should also be given to the health and fitness of the staff as this can have an effect on the safety of the operation. Does the job put at risk someone with a health problem? Does it require someone with unusual strength or height?

Other factors

Personal protective equipment or other clothing should not be used unless absolutely necessary. For example wearing gloves may impair dexterity although they may be required because of cold conditions or to avoid injury from rough or sharp edges. Other protective clothing may make free movement difficult.

Remember

Mass is the quantity of material in a body. The mass of the load is measured in **kilograms** – abbreviated to kg.

The weight of an object is the force which, because of gravity, it exerts on a platform placed under it. **The unit of force is the newton (N).**

When an assessment of the task has been completed a decision must be made as to how the work is to be safely carried out. This may involve more than one person, using special equipment, preparing the working area, using safety clothing etc. If your task requires the use of appropriate equipment and this has been provided by your employer, and you have been trained to use it, then it is a legal requirement for you to do so.

Manual lifting

Whenever a load is to be lifted careful consideration needs to be given to the task. If it is to be lifted manually extra consideration needs to be given to the position of the body.

It is important to keep the spine in its naturally upright position (Figure 8.10). If the back is unnaturally arched forward there is a greater risk of injury to it. The knees should be bent so that when the legs are straightened, the load is lifted. To achieve this the load should be kept close to the body with arms as straight as possible and both hands should be used.

Naturally upright

Figure 8.10

To allow for the centre of gravity of the load the body should lean back slightly when lifting allowing the body to act as a counter balance to the load.

When the load has been lifted and the body is straight care must be taken to avoid sudden turning or twisting as this can also cause damage to the back.

Remember

When lifting and carrying a load
- **STOP** and **THINK**
- place your feet correctly and bend your knees
- keep your back in its naturally upright position
- keep your shoulders level and facing in the same direction as your hips
- keep your elbows in
- hold the load as close to you as possible
- position your hands so that your fingers do not become trapped
- use two hands and get a firm grip
- don't jerk
- put the load down before adjusting the position

Carrying long loads

When a long load is being carried by a single person care must be taken so that injuries do not occur. These injuries may be to the person carrying the long load or somebody else in the vicinity! The person carrying the load should ensure that the centre of gravity is directly related to the carrying position.

In Figure 8.11 the centre of gravity of the ladder is approximately above the shoulder on which it is being carried. To avoid unnecessary injury to others the front end of the load is kept high.

Lifting platforms

When a load has to be lifted for stacking or lifted on to the shoulders to be carried, it is useful to have a lifting platform (Figure 8.12) so that the lift can be done in two stages. This will reduce the risk of strain or injury.

Figure 8.12   Lifting platform.

Pushing and sliding

Lifting a load generally requires more effort than moving a load in a horizontal plane. When considering moving a load, look at all the possible alternatives, as in some cases it is easier to push or slide a load rather than lift it (Figure 8.13). If a load is to be pushed or pulled care must be taken not to damage the operative's back. To work at maximum efficiency with the least possibility of damage to the person, the back should be kept straight and the legs should do the pushing or pulling.

Figure 8.13   Sometimes it is easier to push or slide a load.

Figure 8.11

Points to remember ◀ — — — — — — — — — — — —

There are a number of factors that must be considered when a load has to be moved, which include:

Is it necessary to move the load?

What risks may be encountered with the load itself? (give examples?)

What risks may be encountered on the route? (give examples)

Are you, or is the person who is to carry the load, capable of carrying it?

It is very important when manually lifting a load to consider the position of the body. Damage to the back can be caused by incorrect positioning, sudden turning or _____.

Lifting platforms can reduce the risk of strain or injury by creating another stage in the lifting process.

Lifting a load generally uses more effort than moving a load in a horizontal plane. When pushing or pulling a load, what consideration must be given to the position of the body?

Try this

You are required to lift a bulky load of about 15 kg from the floor and take it to a new position on a worktop 5 m away. Assume the load is on the floor in front of where you are sitting now.

Assess the situation, list any sources of hazard you feel you could encounter and describe how you would deal with them.

Part 2

Assisted moving

Pushing or pulling a load may be easier than lifting, but a heavy load on a flat surface can create a large friction resistance area.

This friction can be reduced if rollers are used between the load and the floor.

Levers

So that rollers can be placed under the load it has to be lifted one end at a time. The lifting can best be carried out with a lever. This is placed under one side of the load and then pushed down (Figures 8.14 and 8.15). As the part of the lever that is pushed down on is a great deal longer than the end that is lifting the load a mechanical advantage is created.

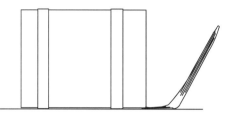

Figure 8.14

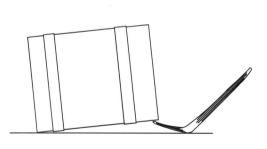

Figure 8.15

The lever may be of the bent type as shown in Figure 8.15 or a straight one used with a block as shown in Figure 8.16.

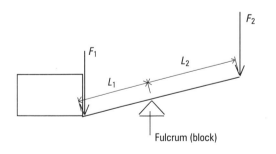

Figure 8.16

The point at which the lever pivots is called the **fulcrum.**

If a lever is the same length on either side of the fulcrum then a force equal to that of the load will create a state of equilibrium. When the force applied to the "handle" of the lever exceeds that of the load, movement takes place and the load is lifted on one edge. To create a mechanical advantage, so that less effort needs to be used, the handle end of the lever is made longer.

A lever with a ratio of 4:1, that is to say, a lever as shown in Figure 8.17 where the handle (L_2) is 4 times longer than the lifting arm (L_1), needs only ¼ the effort to lift the load.

This can be calculated by the fact that to produce equilibrium:

$$F_2 \times L_2 = F_1 \times L_1$$

Consider a load of 10 kg to be lifted by a lever, as shown in Figure 8.17.

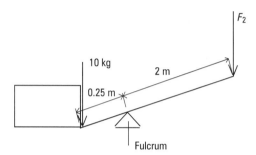

Figure 8.17

As the handle of the lever is 8 times as long as the blade the force required at F_2 has been reduced by 8:1 and therefore the load can be raised using, 10 kg ÷ 8 = 1.25 kg.

Another example of how levers can be used is a pair of pliers or cutters, as shown in Figures 8.18 and 8.19.

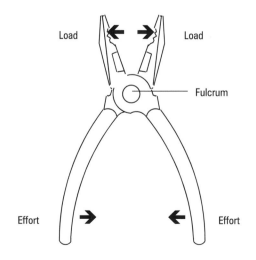

Figure 8.18

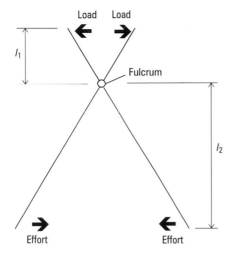

Figure 8.19

Rollers

When the load has been raised rollers can be placed underneath as shown in Figure 8.20. The load can then be pushed gently forward on two or three rollers and further rollers placed under the front end as necessary.

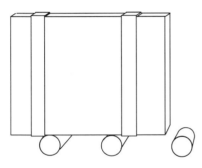

Figure 8.20

Try this

What is the load that can be lifted with an effort of 12 kg on a lever with a distance of 0.3 m between the load and the fulcrum and 0.6 m between the effort and the fulcrum?

Wheelbarrows and trolleys

A **wheelbarrow** (Figure 8.21) has a single front wheel or ball and can be used to carry heavy or bulky loads. It is also useful for carrying loose loads such as sand or gravel. Load the barrow so that it will not overbalance, use both hands, keep a straight back and steady the barrow before moving off.

Figure 8.21   A wheelbarrow.

A **sack barrow** (Figure 8.22) has two wheels and is more stable than a wheelbarrow. It is still very necessary to load the barrow correctly to avoid overbalance.

Figure 8.22   A sackbarrow or hand truck.

A **flat trolley** (Figure 8.23) has four wheels and it is often used in stores where materials are constantly being moved. When on level ground a flat trolley is usually pulled whereas barrows are usually pushed. It may be necessary to prevent trolleys from moving at the wrong time in which case chocks (blocks or wedges) need to be placed to prevent the wheels turning.

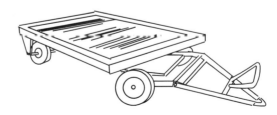

Figure 8.23   A flat trolley.

Fork-lift truck

Another way of moving a load is by using a fork-lift truck. These are often used in large stores where the goods are to be stacked on pallets. Only authorised and trained personnel are allowed to use these.

Wheelstand

A **wheelstand** (Figure 8.24) is a raised platform on two wheels and two feet to avoid the need for lifting the load.

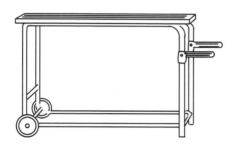

Figure 8.24   *A wheelstand.*

Slings and pulleys

When a heavy load has to be lifted vertically slings and pulleys can be used, but care must always be taken to ensure that the supports are capable of taking the maximum load. The maximum Safe Working Load (SWL; Figure 8.25) should **NEVER** be exceeded.

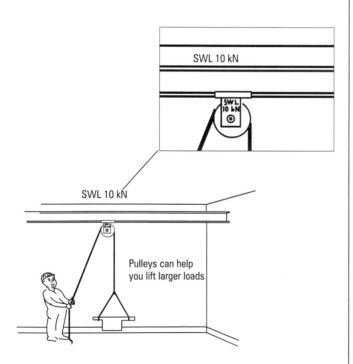

Figure 8.25

After the load has been lifted care must be taken to make sure that nobody can get trapped or crushed by the load (Figure 8.26).

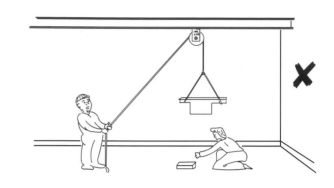

Figure 8.26   *Keep clear!*

NEVER leave a suspended load unsupervised. When you lower a load, do it gently into position and, before you remove the lifting equipment, make sure that the load is stable and will not topple over.

Every load has a centre of gravity. This will not always be the centre of the object. The centre of gravity is determined by the shape and mass at different points.

To ensure even lifts slings are used. Figure 8.27 illustrates how important it is to use slings sensibly as shown in (a), or the load will try to take up a new position as indicated in (b).

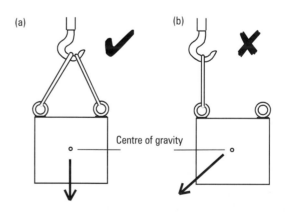

Figure 8.27

Many objects that have to be lifted have sharp edges and corners and rope slings should not be used unless these are protected as they may become damaged (Figure 8.28).

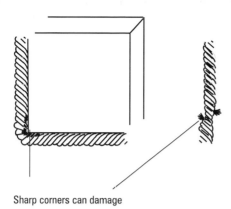

Sharp corners can damage

Figure 8.28

The pulley block

The pulley block consists of a continuous chain or rope passing over a number of pulley wheels as shown in Figure 8.29. Pulley blocks should be regularly tested and their safe working load displayed on them, which should **NEVER** be exceeded.

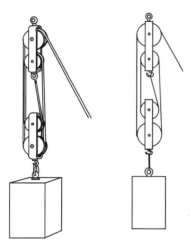

Figure 8.29

When loads are suspended on pulley systems they have a tendency to swing and twist. This problem is often overcome by having a stabilising or control rope tied to the load and manned by a person given the sole responsibility of keeping the load straight.

Loads should never be left suspended in mid-air without someone to watch them. The area under the load should be kept clear at all times in case the load should fall.

When the load has been lowered gently into position it should be checked for movement before the sling is taken away.

Why use pulleys?

Pulleys can give several advantages when trying to lift an object. If a single wheel pulley is used there is no mechanical advantage but a difficult shaped object can be slung so that a single rope can lift it.

Where a pulley system with two pulleys is used the amount of effort required is only half that of a one pulley system. If we go to a four-pulley system like the one shown in Figure 8.29 the effort required is only a quarter (one fourth) of that required with a single pulley.

Example

When lifting a load of mass 16 kg with a single pulley the effort required is also 16 kg.

With 2 pulleys the same load can be lifted with an effort of:

 2 pulleys require half the effort

 i.e. 8 kg

With four pulleys the same load can be lifted with an effort of:

 4 pulleys require one quarter the effort

 i.e. 4 kg.

Try this
1. What mass can be lifted with an effort of 15 kg on a 2 pulley system?

2. On a 4 pulley system a load of 36 kg has to be lifted. What effort is required?

3. A man has to lift a mass of 72 kg. The effort he uses is 18 kg.
 What pulley system does he require in order to lift the load?

Winches

A simple winch, as in Figure 8.30, consisting of a drum around which a rope is wound can also be used to raise loads. A crank handle rotates the drum and takes up or lets out the rope thus raising or lowering the load.

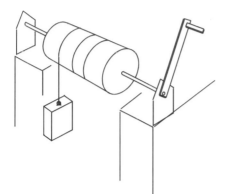

Figure 8.30

Power sources available

Lifting gear can be classified by its power source:
- Mechanical: winches, pulleys, manual (muscle) power – probably yours!
- Electrical: electric motors
- Pneumatic: compressed air
- Hydraulic: liquids
- Petrol or diesel engines: although *not* used in workshops because of the fumes.

Points to remember

Generally a person can carry loads of up to _____ kg. Above that assistance is advisable. This may be extra people or mechanical equipment.

If moving from one place to another there are a number of wheeled devices available (give examples).

To lift loads up to a new height _____ may be used.

Whenever loads are moved care must be taken not to cause an accident or injury.

Self-assessment multi-choice questions
Circle the correct answers in the grid below.

1. Which of the following is correct when manually lifting a load from the floor?
   (a) straight legs and bent back
   (b) upright back and straight legs
   (c) bent legs and load at arms length
   (d) bent legs and upright back
2. A load of 24 kg is to be lifted using a lever and pivots as shown in Figure 8.31. The force that needs to be applied to the end of the lever to lift the load is

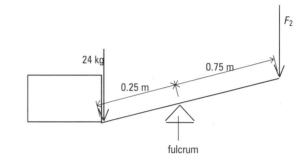

Figure 8.31

   (a) 48 kg
   (b) 24 kg
   (c) 12 kg
   (d) 8 kg
3. The effort required to lift a load of 40 kg when using a four-pulley system is
   (a) 10 kg
   (b) 20 kg
   (c) 40 kg
   (d) 160 kg
4. A wheelstand is used
   (a) to lift a heavy load
   (b) to transport a heavy load to a new position at a similar height
   (c) to transport a heavy load at low level
   (d) in the place of rollers
5. The load generally accepted as being the maximum for a fit person to lift is
   (a) 5 kg
   (b) 15 kg
   (c) 20 kg
   (d) 50 kg

Answer grid

1	a	b	c	d
2	a	b	c	d
3	a	b	c	d
4	a	b	c	d
5	a	b	c	d

9

Access Equipment

From what you learned in the previous chapter can you classify the following lifting devices by their power source?

Match these lifting devices to their power source:
block and tackle, diesel engine, hydraulic lift, electric motor, winch, petrol engine

Power source	Lifting device
manual power	
electrical power	
liquids	
combustion	

On completion of this chapter you should be able to:

◆ identify suitable access equipment for specified situations
◆ recognise the need for suitable staging when using trestles
◆ describe the correct angle for erecting ladders
◆ describe methods of securing ladders
◆ recognise the need for safety requirements when working with tower scaffolds
◆ recognise the need for guardrails and toeboards on scaffolding
◆ complete the revision exercise at the beginning of the following chapter

Figure 9.1   Ladders and other access equipment.

Part 1

Simple access equipment

Access equipment provides a means of reaching the area where work has to be carried out.

Different access equipment will be required depending on the height at which work is to be carried out. Reaching just above shoulder height may only require a step up, whereas scaffolding may be required for access to work on the roof of a building.

There are some basic rules which apply whichever access equipment is required.
- All access equipment should be set up on a firm level base.
- The equipment chosen must be suitable for the task so that the user does not have to over-reach.
- All access equipment should be inspected regularly to ensure that it is in good condition. This does not just mean whether or not it is broken, but also looking to see if the surface is slippery because of mud or ice, or other similar hazard.

Step-up

To reach up a short distance in comfort use a very simple piece of equipment called a step-up, often referred to as a hop-up (Figure 9.2).

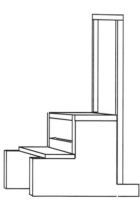

Figure 9.2   A step-up

Steps

For work up to the height of a ceiling a pair of steps (Figure 9.3) could be used. These should be high enough for the user to stand with his knees below the top of the steps, which should be open to their fullest extent and should be set on level and firm flooring.

The knees should be below the top of the steps

Figure 9.3   A pair of steps.

Trestles and platform

Some work at ceiling height involves working over some distance and in such cases trestles and a platform may be used (Figure 9.4).

Open the trestles to their full extent and if stay bars are fitted lock them into place.

Figure 9.4

The platform can be either two scaffold planks wide (at least 450 mm) or a lightweight staging and it should be placed no higher than two thirds of the way up the trestles.

The platform must not overhang the trestles by more than four times the thickness of the boards (Figure 9.5) and the minimum overhang allowed is 51 mm. So if two scaffold planks are used, each 50 mm thick, then they must not overhang by more than

$$4 \times 50 = 200 \text{ mm}$$

Due to the extent of injuries that could result from falling the maximum height for the platform must not exceed **4.57 m**.

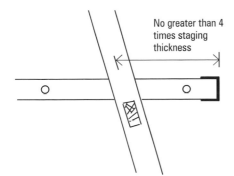

Figure 9.5

If the platform is over 2 m high a pair of steps should be used to gain access.

The platform and trestles should always be dismantled before moving them to a new position.

Ladders

If access to somewhere higher is required then a ladder should be used. The parts of a ladder are shown in Figure 9.6.

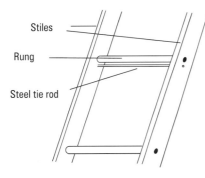

Stiles

Rung

Steel tie rod

Figure 9.6

It is very important when working on a ladder to remember the rule about standing it on firm level ground, and if it has to be put on soft earth then use a board (Figure 9.7). Never use boxes, drums or any other unsteady or sloping base.

Figure 9.7

There are also a number of other points to take note of:

- Ladders must not be too short as this could cause the user to over-reach and fall off.
- Ladders must extend 5 rungs or 1.05 m above the working platform unless there is an adequate handhold to reduce the risk of overbalancing (Figure 9.8).

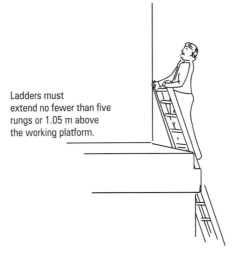

Ladders must extend no fewer than five rungs or 1.05 m above the working platform.

Figure 9.8

- Ladders must be inspected frequently to ensure they are in good working order and that no rungs are damaged, missing or slippery. Damaged ladders should be withdrawn from use, and clearly labelled that they are damaged and not to be used.

> ### Try this
>
> What would be the most appropriate access equipment to reach the ceiling of the room you are in?
>
>
> What would be the most appropriate access equipment to reach a height of 5 m?
>
>
> What would be the most appropriate access equipment to reach a height just above your own?

What is the highest position a platform can be placed up the trestles?

What is the maximum height for a platform?

If the platform is over 2 m high how should access be gained?

How far above the working platform must a ladder extend?

Part 2

Extension ladders and scaffolding

When extending a ladder there must be an overlap of rungs:

Figure 9.9

Two rungs for ladders with a closed length of up to 5 m (Figure 9.9).

Figure 9.10

Three rungs for ladders with a closed length of up to 6 m (Figure 9.10).

Figure 9.11

Four rungs for ladders with a closed length of over 6 m (Figure 9.11).

When moving ladders any distance they should be carried on the shoulders of two people, one at each end of the ladder (Figure 9.12).

Figure 9.12   Two people are required to move long ladders.

Raising ladders

Ladders should be raised with the sections closed. Two people may be required to raise the heavier ladders.

One stands on the bottom rung and holds the stiles to steady the ladder while the second one stands at the top and raises the ladder above head height and walks towards the bottom moving his or her hands down the ladder while walking (Figure 9.13).

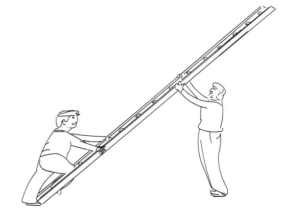

Figure 9.13

When the ladder is in position it must follow the 4:1 rule (Figure 9.14). This means that the height of the ladder above the ground must be four times the distance the ladder is out from the foot of the wall. The ladder will then be at an angle of 75° to the ground.

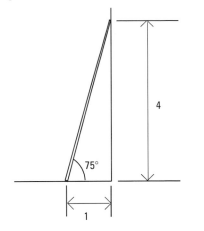

Figure 9.14

If the ladder is over 3 m long then it must be secured or another person must "foot" the ladder. To secure the ladder it must be lashed to a secure position like a scaffold pole (Figure 9.15). A drainpipe or gutter is not a secure position and should not be used.

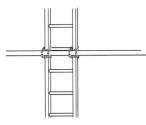

Figure 9.15

In some cases it is necessary to secure ladders of less than 3 m, as even short falls can cause injuries.

Figure 9.16

To foot the ladder another person must stand with one foot on the bottom rung, one foot on the ground and both hands holding the stiles (Figure 9.16). The purpose of footing the ladder is to prevent it moving, and the person carrying out this task must remain alert and observant at all times.

Example

An extending ladder is at an angle of 75° with the ground (Figure 9.17). How high up the wall will the ladder be resting if it extends 1.5 m out from the base of the wall?

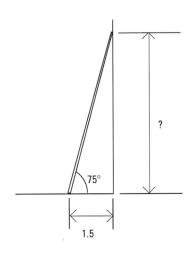

Figure 9.17

Answer

If the ladder is at the angle of 75° then the 4:1 rule applies. This means that the height of the ladder above the ground is 4 times 1.5 m.

$$4 \times 1.5 = 6 \text{ m}$$

Try this

How far out from the base of a wall should an extending ladder be if the top end of the ladder rests on the wall at a height of 5.6 m (Figure 9.18)?

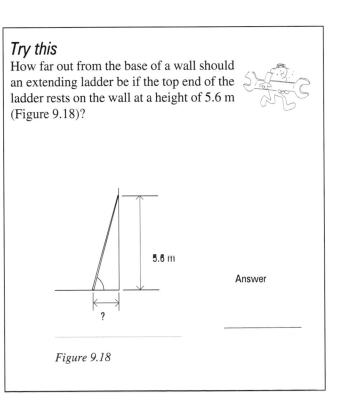

Answer

Figure 9.18

Tower scaffold

For working at, and not just gaining access to, these higher levels a tower scaffold can be used (Figure 9.19).

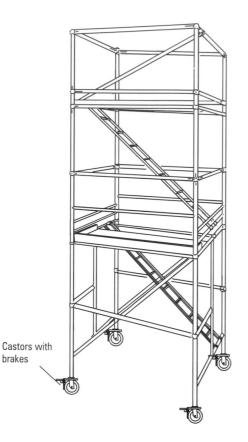

Castors with brakes

Figure 9.19   Tower scaffold.

Training in the erection, use and inspection of the tower scaffold should be given. Operatives who have not been given this training should not erect or use the scaffold without suitable supervision.

The minimum base measurement for any tower is 1.21 m and for outside use the height must *not* exceed 3 times the smallest base measurement.

If it is higher than 4 m outriggers must be fitted, and if it exceeds 9 m it must be anchored to the ground or tied to the building.

The overall height of the tower must not exceed 12 metres unless it has been specially designed to do so.

The castors at the foot of the tower must be locked before anyone climbs the tower, and if someone should be working on the tower it must not be moved. When the tower has to be moved it should only be moved by pushing at the base.

Any platforms used for working from must have toe-boards and guardrails for the safety of those working on and those below (Figure 9.20). Precautions must also be taken to ensure that access cannot be gained to incomplete scaffold towers and that only authorised persons can gain access at any time.

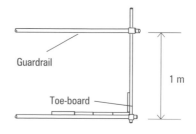

Guardrail

Toe-board

1 m

Figure 9.20

Where safety equipment, for example a harness (Figure 9.21), is used on tower scaffold it must be designed and constructed to prevent serious injury in the event of a fall.

Figure 9.21   A harness used on tower scaffold.

Scaffolding

Full scaffolding is another alternative, but this should only be erected by a skilled scaffolder.

Choice of access equipment will depend on the use for which it is required.

For just a vertical height use a step-up, steps or a _____.
For both horizontal and vertical heights use either trestles and a platform or some form of scaffold.

With trestles and platform remember the overhang rule (_____) and the highest distance that the platform should be placed _____. With a ladder remember the 4:1 rule to set it at the correct angle (_____) and the _____ rungs that must extend above the working platform.

All access equipment needs to be set up on a firm level base, suitable for the task and to be in good condition.

Aluminium ladders are available but should not be used near electrical equipment or an electrical supply because of the possibility of causing a short circuit and receiving an electrical shock.

Try this

You were on site when the following situation occurred and have been asked to make a report to your company. Write the report giving as many details as you can regarding the problem so that it will not happen again.

An electrician was working at a height which required the use of a ladder and he requested that his mate foot the ladder. Whilst the electrician carried out his work his mate was distracted and let go of the ladder. The ladder slipped and the electrician fell to the ground.

Put in details such as the type of work involved, the distraction and the extent of the injuries sustained by the electrician.

Self-assessment multi-choice questions

Circle the correct answers in the grid below.

1. The minimum number of rungs of a ladder that must extend past the working platform is
    (a) 1
    (b) 3
    (c) 5
    (d) 7

2. When trestles and platform are used, the overhang of the platform should not exceed the thickness of the platform multiplied by
    (a) 2
    (b) 4
    (c) 6
    (d) 8

3. When a ladder is leant against a wall the ratio of the height up the wall to the distance away from the wall at the bottom should not exceed a ratio of
    (a) 2:1
    (b) 4:2
    (c) 4:1
    (d) 5:3

4. The maximum height a tower scaffold can be used without outriggers is
    (a) 3 m
    (b) 4 m
    (c) 5 m
    (d) 6 m

5. When an extending ladder with a closed length of 3 m is fully extended the minimum number of rungs overlap must be
    (a) 1
    (b) 2
    (c) 3
    (d) 4

Answer grid

1	a	b	c	d
2	a	b	c	d
3	a	b	c	d
4	a	b	c	d
5	a	b	c	d

10

Commissioning the Installation

You will need to have a copy of IEE Guidance Note 3 available for reference in order to complete some of the exercises within this chapter. If you can gain access to a copy of the IEE Wiring Regulations (BS7671) it would be helpful to read through Part 7.

Check that you can remember the following from Chapter 9.

All access equipment should be set up on

The equipment chosen must be suitable to the task. The user should not have to

All access equipment should be inspected regularly to ensure that it is in good condition. What should you look for?

When you have to work at ceiling height over a long distance it would be appropriate to use

Ladders must extend _____m above a working platform.

Draw a sketch of a ladder and label the stiles, the rungs and the steel tie rod.

How could you secure a ladder of over 3 m long? (two answers)

Figure 10.1   Testing equipment.

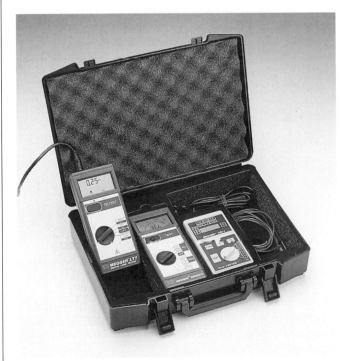

On completion of this section you should be able to:

◆ recognise the need for inspection and testing of an installation
◆ be familiar with the instruments used in this process
◆ recognise the need to isolate and secure an installation before testing
◆ identify safe working practices for visual inspection
◆ list the appropriate tests and identify test equipment required
◆ complete the end test contained in the back of this book

Figure 10.2   16th Edition test kit.

Part 1

Reasons for inspection and testing

Inspection, and where necessary, testing, should be carried out during all stages of erection and before the electrical installation is put into service. This is to verify that the installation is safe, conforms to the relevant regulations, codes of practice and installation specification, and is suitable for its purpose.

When this has been carried out the installation can be commissioned, which demonstrates that everything operates properly and safely.

Certain tests must be completed before the installation is connected to the supply and energised.

Existing installations require testing periodically and may also require testing on change of ownership and on change of use of the building (Figure 10.3).

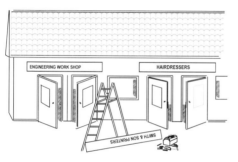

Figure 10.3

In the same way that a vehicle technician needs manuals and documentation when servicing a car (Figure 10.4), documentation and publications are required during the inspection and testing of an installation (Figure 10.5).

Figure 10.4   Documentation is required when servicing a car.

The documentation requirements for inspection and testing are described in Part 7 of BS 7671, The requirements for electrical installations, which should be used in conjunction with the IEE Guidance Note 3 on this topic.

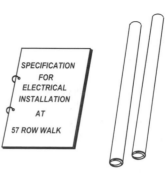

Figure 10.5   Documentation is required when commissioning an installation.

The documentation required includes the specifications, design data, drawings and commissioning procedures associated with the installation. This information is often in the form of an operational manual or schedule which will enable a competent person to carry out this work.

A competent person will be skilled, experienced and have sufficient knowledge to ensure that no danger occurs to any person, livestock or property during the process of inspection and testing.

Safety during inspection and testing

Whilst inspection and testing is carried out there must be no risk of danger to persons or risk of damage to equipment or property.

Before any visual inspection is undertaken it is vital to ensure that the installation is completely isolated from the supply and locked off to prevent the possibility of accidental reconnection (Figure 10.6). This is important because during the inspection it is necessary to look thoroughly into equipment to examine parts which would normally be live.

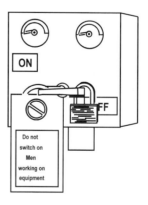

Figure 10.6

Testing to ensure that the installation has been isolated must only be carried out using an approved test instrument which must be "proved" before and after making the test using a known supply.

Also:

- notices should be displayed at the entrances to buildings and at the intake point to warn others that testing is being carried out.
- no one other than the persons carrying out the inspection or testing may be working on the installation under test
- the installation must not be used during the test

It will be necessary to liaise with other contractors and persons involved on the site before commencing this work, as temporary supplies could be cut off and alarm systems initiated.

Remember

- Understand the equipment you are proposing to use
- Understand the application to which you intend putting it
- Ensure that safety procedures are followed
- Do not take chances or shortcuts

The visual inspection

The purpose of the initial inspection is to verify as far as possible that

- all equipment and materials are of the correct type and meet appropriate standards (Figure 10.7)
- all parts of the installation are correctly selected and erected
- no parts are visibly damaged or are defective
- the installation is suitable for its purpose
- the installation conforms to specifications, and complies with current regulations, codes of practice and guidelines (HASAWA, EAWR, BS 7671 and the IEE Guidance Notes)

BS 7671 includes a checklist for inclusion in the visual examination.

The visual inspection of the installation requires examination of the environmental conditions in which it is to operate, the use any building is to be put, and the type of wiring system and the load likely to be imposed on it.

Figure 10.7   Verify that equipment is of the correct type.

Try this
From IEE Guidance Note 3 make your own checklist of items for visual examination.

99

During the inspection we will be looking to make sure that the type of installation is suitable for its purpose (Figure 10.8). For example, a wiring system serving damp farm buildings should be suitable for the environment and fitted out of the way of animals.

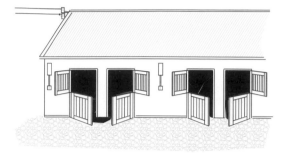

Figure 10.8   *The visual inspection ensures that the type of installation is suitable for its purpose.*

In an industrial situation you could expect to find PVC wire armoured cables fixed to galvanised tray away from the risk of mechanical damage.

For less demanding situations such as in the roof space of a house we may expect to find PVC/PVC cables neatly clipped to the side of a joist above the level of the loft insulation (Figure 10.9).

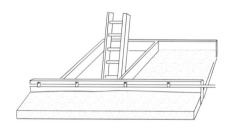

Figure 10.9   *Cables clipped above the loft insulation.*

In an office it is likely that cable management systems such as compartmentalised dado or skirting trunking would be used (Figures 10.10 and 10.11).

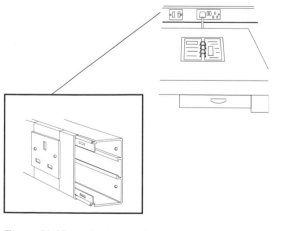

Figure 10.10   *Dado trunking.*

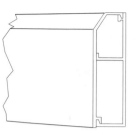

Figure 10.11   *Skirting trunking.*

During inspection we should be on the lookout for anything which may reduce the integrity of the wiring system. For example, the proximity of pvc cables to a heat source, dampness where black enamelled conduit has been used, unprotected cables and so on.

In addition to ensuring that the installation is fitted in accordance with the original design, the inspection should identify possible changes that have occurred between the time of the original design and its commissioning. The building may change owner and use, so that the new positioning of equipment and loadings mean that cables are no longer adequately sized. Alteration to pipework may affect the integrity of the equipotential zone and repositioned walls may reduce the effectiveness of fire barriers where wiring systems pass through walls or floors and so on.

Try this

Your company offers a wiring inspection service to customers of a local estate agent when they purchase a house. You have been asked to write to some new customers regarding what the service entails. The customers are purchasing an old property in a local village.

100

Commissioning the installation is done to demonstrate that
_____.

The documentation requirement for inspecting and testing
installations includes

Safety is of vital importance when undertaking an inspection.
What should be done to ensure that there is no risk of danger
to persons or damage to equipment or property?

Remember

A detailed inspection of an installation will
confirm that it is suitable for its occupant,
usage and the environmental conditions
including the following types:

agricultural premises

commercial premises

domestic properties

industrial sites

Part 2

Testing

Inspection and testing should be carried out on all new works
and when any alterations are made.

In addition all installations deteriorate due to wear, corrosion,
overload and environmental influences, and so should be tested
periodically, preferably at not less than the intervals set out in
the IEE Guidance Notes 3, Inspection & Testing. Some
examples of the frequency are shown in the table below:

Table 6.1

Reproduced from Guidance Note 3 by kind permission of the Institution of
Electrical Engineers.

Type of building	Maximum period between inspections
Cinemas	1 year
Petrol filling stations	1 year
Domestic	10 years
Agricultural & Horticultural	3 years
Construction site installations	3 months

Periodic testing of an installation, for example in a factory or
shop, is likely to cause disruption of business, so the timing of
the testing and the extent of the shutdown will need to be
established, preferably in a written specification.

The following information should be made available to the
person carrying out the inspection and test:
- Diagrams, charts and tables indicating

  the type of circuit
  the number of points installed
  the number and size of conductors
  the type of wiring system

- The location and types of device used for

  protection
  isolation and switching

- Details of protective devices used, the earthing arrangements, the impedances of the circuits and a description of
  how automatic disconnection of the supply is achieved in
  the event of a fault.

All of this information is likely to be included in the contract
specification, together with manufacturers' data.

Instruments required for installation testing

The following instruments are required to test a simple installation:

a low reading ohmmeter
a 500 volt megohmmeter
} NOTE these are usually combined in one instrument.

an earth fault loop impedance meter.

An RCD test instrument and an earth electrode test instrument will be required for some installations (Figure 10.12).

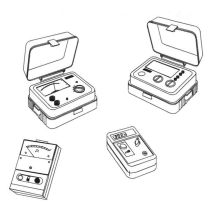

Figure 10.12

Good practice, and the requirements of the Electricity at Work Regulastions 1989, require test instruments to be regularly checked and recalibrated.

The serial number of the instrument should be recorded with the test results obtained for the installation.

Any defects, damage, omissions or signs of deterioration found during the inspection and test must be recorded, preferably in writing, and passed to the appropriate person together with the test results.

The test results

When the inspection and test has been completed, the results must be recorded on the form prescribed in BS 7671 and forwarded to the person ordering the report. Details of the tests required will be found in the book "Procedures".

The record will include any dangerous conditions arising through non-compliance with BS 7671 and any limitations of the inspection and testing. If any serious defects are noted these should be reported immediately and action taken to minimise danger.

Try this

You have been given the task of explaining to a new apprentice what precautions should be taken to ensure the safety of himself and others before using test instruments. Using BS 7671 and IEE Guidance Note 3 as a reference make a note, in your own words, of the precautions listed.

Remember

Details of the following tests can be found in the book "Procedures".

Tests to be undertaken (and in this order):

continuity of protective conductors, main and supplementary bonding

continuity of ring final circuit conductor

insulation resistance

Fault finding and diagnosis

It is important to know how to find out if there is a fault and what to do about it to correct it.

Procedures for fault finding follow a logical progression and proceed step-by-step.

Problem

Obtain information available

Analyse information

Note the options

Choose the most appropriate

Did it solve the problem?

(If no, analyse the results of the test
and either choose another option or,
if more appropriate,
obtain more information and start again.)

Let's look at the steps in more detail and apply them in particular to fault finding and diagnosis in electrical work.

Obtain information available
The information will include such factors as
- knowledge and understanding of the plant, equipment and system concerned
- personal and others' experience and expertise
- data regarding the fault (both the occurrence and leading up to it) assembled from both verbal and written reports and first-hand experience if possible.

Analyse information
Collate the information and list any appropriate actions that may be considered relevant.

Note the options
There may be one or more than one option available, so it may be appropriate to use fault location techniques and/or basic tests to diagnose the cause (or causes) of the fault.

Choose the most appropriate option
After having chosen the most appropriate course of action carry it out and interpret the results. If the fault has been located, can it be repaired or does the part need to be replaced?

Did it solve the problem?
Checks will have to be made to ensure that the fault has been rectified correctly.

Example:
An electric fan heater is found not to be working correctly.

Information available:

> Although the fan is working no heat is being given out by the fan heater.

> Previously a sheet airing nearby had been knocked from the airer and had landed on the fan heater, covering it.

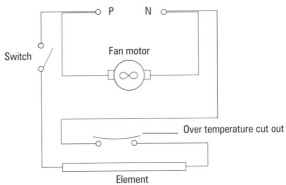

Figure 10.13

The circuit diagram for the fan heater is available (Figure 10.13).

Analyse information:

> If the fan is working correctly then the supply must be reaching the appliance.

> In the heater circuit there is the on/off switch, an over temperature cut out and the heater element.

Note the options:

> A fault with one or more of the following could be the cause of the heater not working:

> the on/off switch
> the over temperature cut out
> the heater element

Choose the most appropriate option:

As a sheet had covered the heater the air supply may have been cut off and the element overheated. This could mean the over temperature cut out needs resetting or that the element has gone open circuit.

If the over temperature cut out has worked correctly this should be the problem.

Did it solve the problem?

Although the temperature cut out should be the most obvious cause, if it turns out not to be then the second option should be investigated.

If the temperature cut out has proved to be the problem some damage may have been caused due to the overheating. To ensure the heater is safe to use, further tests should be carried out (Figure 10.14).

Figure 10.14   Use recognised test apparatus to carry out tests on equipment after repair.

Points to remember ◀ – – – – – – – – – – – – – –

A visual inspection and tests verify, as far as possible, that acceptable standards have been achieved. In order to do this, information regarding the installation must be available as well as current British Standards, codes of _____ and guidelines. Before any visual inspection is undertaken the installation should be _____ and precautions taken to prevent accidental reconnection.

Inspection and testing should be carried out on all new installations, both during and on completion, at recommended intervals after completion and when any _____ or _____ are made to the installation.

When tests are carried out the results should be recorded and checked against the minimum acceptable values, and if discrepancies occur these need to be examined and corrected.

Circle the correct answers in the grid below.
1.   The purpose of inspecting an installation is to
      (a) calculate the overall cost of the installation
      (b) make a report on the progress of the contract
      (c) confirm whether the installation meets appropriate standards
      (d) ascertain whether further materials will be required
2.   The first action to be taken when undertaking fault finding is to
      (a) inspect the equipment
      (b) dismantle and check
      (c) obtain all the available information
      (d) refer to manufacturer's details
3.   In order to carry out fault diagnosis an individual must
      (i) have knowledge and experience of the equipment and system involved
      (ii) be able to analyse the information available regarding operation and fault development
      (a) Both statements (i) and (ii) are correct
      (b) Neither statement (i) or (ii) is correct
      (c) Only statement (i) is correct
      (d) Only statement (ii) is correct
4.   It is recommended that domestic installations are inspected and tested at least every
      (a) 1 year
      (b) 2 years
      (c) 5 years
      (d) 10 years
5.   The first of the tests to be undertaken after the inspection has been completed satisfactorily is
      (a) polarity
      (b) insulation resistance
      (c) continuity of ring final circuit conductors
      (d) continuity of protective conductors

Answer grid

1   a   b   c   d
2   a   b   c   d
3   a   b   c   d
4   a   b   c   d
5   a   b   c   d

Try this

Your company is about to offer customers a new service for testing portable appliances.

Design a leaflet which gives details of this new service, giving examples of the type of portable appliance it is appropriate for.

End test

Circle the correct answers in the grid at the end of the multi-choice questions.

1. Nuclear power stations use which process to produce electricity?
   (a) tidal flow
   (b) geothermal
   (c) direct drive
   (d) steam

2. When extra demand for electricity is required for a short period only, which of the following is most likely to be used?
   (a) hydroelectric power
   (b) wind generator
   (c) gas turbine generator
   (d) coal-fired power

3. The supply voltage from the Electricity Companies must be within what percentage of the stated nominal voltage?
   (a) +6%, –10%%
   (b) +10%, –4%
   (c) +10%, –6%
   (d) +4%, –10%

4. Area Electricity Companies generally take their supplies at
   (a) 132 kV
   (b) 230 kV
   (c) 400 kV
   (d) 33 kV

5. The consumer's installation starts after the
   (a) circuit protection but before the consumer's load
   (b) meter but before the consumer's isolator
   (c) main supply fuse but before the meter
   (d) consumer's isolator but before the circuit protection

6. Another term for an energy meter is a
   (a) current meter
   (b) voltage meter
   (c) kilowatt hour meter
   (d) ohmmeter

7. The person on site who would normally prepare the "as fitted" drawing would be the
   (a) electrical subcontractor's site foreman
   (b) architect
   (c) client
   (d) main contractor

8. Electrical equipment used out of doors should have an RCD protecting the circuit. If a fault to earth develops the circuit should automatically switch off before the fault to earth current reaches a maximum of
   (a) 30 mA
   (b) 40 mA
   (c) 50 mA
   (d) 60 mA

9. The BS EN 60309-2 colour of plug for a voltage of 110 V is
   (a) white
   (b) yellow
   (c) blue
   (d) red

10. Which of the following would be suitable to put out an electrical fire in a computer?
    (a) fire blanket
    (b) water fire extinguisher
    (c) foam fire extinguisher
    (d) CO_2 gas fire extinguisher

11. A health and safety policy is
    (a) a safety guideline produced by the government which must be followed in all places of work.
    (b) a set of safety rules which you must carry on you and follow whilst at work
    (c) a list of safety rules that your company must follow
    (d) a safety document written by a company which lists its safety rules and who is responsible for safety

12. What category of safety sign means stop or do not do?
    (a) mandatory
    (b) prohibition
    (c) warning
    (d) safe condition

13. What colour and shape are mandatory signs?
    (a) yellow and triangular
    (b) green and square
    (c) red and circular
    (d) blue and circular

14. A blue circular sign with a picture of a person wearing a helmet means?
    (a) WARNING – falling objects may injure your head
    (b) you must wear a hard hat
    (c) only people on building sites must wear hard hats
    (d) you must not wear a hard hat

15. Fire barriers must be replaced when a conduit run
    (a) passes through a structural brick wall
    (b) enters an inspection elbow
    (c) enters a light fitting outlet box
    (d) enters a restricted space

16. The fixing device shown is a

    (a) toggle fastening
    (b) chemical anchor
    (c) clamp fitting
    (d) coach screw

17. Clamp fittings are generally used on
    (a) plasterboard
    (b) chipboard
    (c) steel girders
    (d) bricks
18. Which type of screw is tightened with a spanner?
    (a) common screw
    (b) countersunk screw
    (c) twin thread screw
    (d) coach screw
19. The fixing device shown is

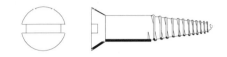

    (a) a hammer screw
    (b) a countersunk screw
    (c) a coach screw
    (d) an expanding anchor
20. A scale of 1:10 is used on a drawing. This means that 1 cm represents
    (a) 100 mm
    (b) 10 mm
    (c) 1 m
    (d) 10 m
21. Which of the following measuring equipment would you use to accurately measure the inside diameter of a pipe?
    (a) a cable gauge
    (b) outside callipers
    (c) inside callipers
    (d) a tape
22. For standard office use a rule is generally of length
    (a) 3 m
    (b) 1 m
    (c) 30 cm
    (d) 30 mm
23. In order to lift a load on to the shoulders which of the following would be most suitable for assistance?
    (a) a lever
    (b) a lifting platform
    (c) a flat trolley
    (d) a hand truck
24. What mass can be lifted with an effort of 18 kg on a 2 pulley system?
    (a) 9 kg
    (b) 18 kg
    (c) 36 kg
    (d) 72 kg
25. What is the load that can be lifted with an effort of 18 kg on a lever with a distance of 0.2 m between the load and the fulcrum and 0.6 m between the effort and the fulcrum as shown?

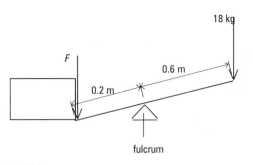

    (a) 54 kg
    (b) 6 kg
    (c) 108 kg
    (d) 27 kg
26. Which of the following power sources should not be used in an inside workshop?
    (a) electrical
    (b) pneumatic
    (c) diesel engine
    (d) hydraulic
27. The symbol shown is the symbol for

    (a) a 2 way switch
    (b) an intermediate switch
    (c) a lighting point
    (d) the main control point
28. The symbol shown is the symbol for

    (a) fluorescent lamp – single tube
    (b) socket outlet
    (c) one way switch
    (d) main control point
29. A plumb line can be used to
    (a) mark out horizontal lines
    (b) transfer a mark on the floor to the ceiling above
    (c) take measurements from a drawing
    (d) mark out a measurement on a drawing
30. The type of drawing which indicates the particular location of outlets and accessories and is drawn to scale is known as
    (a) a circuit diagram
    (b) a block diagram
    (c) a wiring diagram
    (d) a layout diagram

31. An as fitted diagram is usually drawn on copies of
    (a) a site layout drawing
    (b) a block diagram
    (c) the architect's instructions
    (d) a front elevation

32. The maximum testing period between inspection and tests on a petrol filling station is
    (a) 3 months
    (b) 1 year
    (c) 3 years
    (d) 5 years

33. The maximum safe height for a platform resting on trestles is
    (a) 2.45 m
    (b) 3.58 m
    (c) 4.57 m
    (d) 5.65 m

34. How far out from the base of a wall should an extending ladder be if the top end of the ladder rests on the wall at a height of 4.8 m?
    (a) 2.4 m
    (b) 2.0 m
    (c) 1.2 m
    (d) 0.8 m

35. A ladder that follows the 4:1 rule will be at an angle to the ground of
    (a) 45°
    (b) 55°
    (c) 65°
    (d) 75°

36. Ladders must extend no fewer than 5 rungs above the working platform or a distance of
    (a) 97 cm
    (b) 1.05 m
    (c) 1.12 m
    (d) 1.23 m

37. The inspection of an installation should
    (i) Verify as far as possible that no parts are visibly damaged or are defective
    (ii) Confirm that the overall cost of the materials is correct.
    (a) Only statement (i) is correct.
    (b) Only statement (ii) is correct.
    (c) Both statements are correct.
    (d) Neither statement is correct.

38. Where can you find a table which gives you the minimum values of insulation resistance?
    (a) HASAWA
    (b) EAWR
    (c) BS 7671:1992
    (d) On a variation order

39. Which of the following instruments is required to test a simple installation?
    (a) an earth fault loop impedance tester
    (b) a component tester
    (c) a power disturbance analyser
    (d) a digital thermometer

40. Which of the following is unlikely to require special care for disposal?
    (a) fluorescent light fittings
    (b) conduit
    (c) asbestos lagging
    (d) carcinogenic oils

Answer grid

1	a	b	c	d		21	a	b	c	d
2	a	b	c	d		22	a	b	c	d
3	a	b	c	d		23	a	b	c	d
4	a	b	c	d		24	a	b	c	d
5	a	b	c	d		25	a	b	c	d
6	a	b	c	d		26	a	b	c	d
7	a	b	c	d		27	a	b	c	d
8	a	b	c	d		28	a	b	c	d
9	a	b	c	d		29	a	b	c	d
10	a	b	c	d		30	a	b	c	d
11	a	b	c	d		31	a	b	c	d
12	a	b	c	d		32	a	b	c	d
13	a	b	c	d		33	a	b	c	d
14	a	b	c	d		34	a	b	c	d
15	a	b	c	d		35	a	b	c	d
16	a	b	c	d		36	a	b	c	d
17	a	b	c	d		37	a	b	c	d
18	a	b	c	d		38	a	b	c	d
19	a	b	c	d		39	a	b	c	d
20	a	b	c	d		40	a	b	c	d

Answers

These answers are given for guidance and in some instances are not necessarily the only possible solutions.

Chapter 1
p.10 Try this: (1) a: 253 V, b: 216.2 V; (2) a: 50.5 Hz, b: 49.5 Hz
p.12 SAQ (1) c; (2) a; (3) d; (4) a; (5) a

Chapter 2
p.26 SAQ (1) d; (2) c; (3) a; (4) b; (5) c

Chapter 3
p.29 Try this: Hand washing facilities, first aid equipment, protective eye wear (where required)
p.42 SAQ (1) c; (2) c; (3) a; (4) b; (5) a; (6) c; (7) d; (8) b; (9) a; (10) d

Chapter 4
p.50 SAQ (1) d; (2) b; (3) a; (4) d; (5) c

Chapter 5
p.58 (1) c; (2) c; (3) a; (4) d; (5) b
pp.59 and 60, Progress Check:
(1) c; (2) d; (3) b; (4) a; (5) b; (6) c; (7) c; (8) b; (9) c; (10) c; (11) d; (12) a; (13) d; (14) a; (15) c; (16) b; (17) d; (18) d; (19) d; (20) c

Chapter 6
p.68 SAQ (1) a; (2) a; (3) d; (4) d; (5) c

Chapter 7
p.75 Try this:

1 cm	0.5 m
5 cm	2.5 m
25 cm	12.5 m
3 cm	1.5 m
6.5 cm	3.25 m
300 mm	3 cm
1 m	10 cm
20 cm	2 cm
2.8 m	28 cm

p.78 SAQ (1) a; (2) d; (3) c; (4) b; (5) c

Chapter 8
p.81 Try this: 24 kg
p.87 Try this: (1) 30 kg; (2) 9 kg; (3) 4 pulley system
p.88 SAQ (1) d; (2) d; (3) a; (4) b; (5) c

Chapter 9
p.93 Try this: 1.4 m
p.96 SAQ (1) c; (2) b; (3) c; (4) b; (5) b

Chapter 10
p.103 SAQ (1) c; (2) c; (3) a; (4) d; (5) d
pp.105, 106 and 107, End Test:
(1) d; (2) c; (3) c; (4) a; (5) b; (6) c; (7) a; (8) a; (9) b; (10) d; (11) d; (12) b; (13) d; (14) b; (15) a; (16) a; (17) c; (18) d; (19) b; (20) a; (21) c; (22) c; (23) b; (24) c; (25) a; (26) c; (27) a; (28) d; (29) b; (30) d; (31) a; (32) b; (33) c; (34) c; (35) d; (36) b; (37) a; (38) c; (39) a; (40) b